The Geology of the Everglades and Adjacent Areas

The Geology of the Everglades and Adjacent Areas

Edward J. Petuch • Charles E. Roberts

Department of Geosciences, Florida Atlantic University

Photography by Mardie Drolshagen Banks

CRC Press
Taylor & Francis Group
Boca Raton London New York

CRC Press is an imprint of the
Taylor & Francis Group, an **informa** business

CRC Press
Taylor & Francis Group
6000 Broken Sound Parkway NW, Suite 300
Boca Raton, FL 33487-2742

First issued in paperback 2019

ISBN-13: 978-1-4200-4558-1 (hbk)
ISBN-13: 978-0-367-38921-5 (pbk)

Library of Congress Cataloging-in-Publication Data

Petuch, Edward J.
    The geology of the everglades and adjacent areas / author(s), Edward J. Petuch, Charles Roberts.
      p. cm.
    Includes bibliographical references and index.
    ISBN 1-4200-4558-X (alk. paper)
    1. Geology--Florida--Everglades Region. 2. Geomorphology--Florida--Everglades Region. I. Roberts, Charles, 1953- II. Title.

QE100.E9P48 2007
557.59'39--dc22
                                        2006102920

Visit the Taylor & Francis Web site at
http://www.taylorandfrancis.com

and the CRC Press Web site at
http://www.crcpress.com

Dedication

To Our Wives,

Linda J. Petuch and
Susan R. Roberts,

and to our children,

Eric, Brian, and
Jennifer Petuch, Emily
and Sean Roberts, and
Rachel Williams

Foreword

The Florida Platform, and its associated subsurface geology and surface landforms, is clearly a child of the sea. The geologic formations from the latter part of the Mesozoic through the Cenozoic were laid down in shallow seas and periodically reworked by marine, near-shore, and fluvial erosion/deposition processes numerous times. The subsurface lithologies are complex and difficult to correlate due to limited data availability and lithologic similarity. Information is compiled in the form of water well cuttings, stratigraphic core tests, geophysical wire line logs, seismic reflection data, groundwater geochemistry, and limited surface exposures. Further, it is painstakingly difficult to understand and decipher the environmental changes that occurred throughout the geologic past, as the platform grew and was built by the various sedimentary processes.

The authors of *The Geology of the Everglades and Adjacent Areas* have done an outstanding job of compiling decades of data collected by their own field reconnaissance and other geoscientists. They have described the Cenozoic rock sequence units, both formally and informally, to establish the overall lithologic character, paleontology, age, and map extent of the various units. Their sequential paleoenvironmental interpretations and graphical figure representations are a significant contribution to the chronologic understanding of the development of the Florida Platform. They are to be congratulated for the innovative use of modern satellite technology and digital graphics to present their sequential paleogeography interpretations. This represents a significant contribution to the understanding of the development of the Florida carbonate platform, and it will assist other disciplines as they strive for better understanding of our groundwater resources, aquifer characterizations, paleoenvironmental interpretations, and historical/educational geology programs.

Walt Schmidt, Ph.D., PG #1
Florida State Geologist & Chief
Florida Geological Survey

Preface

There is but one Everglades on this planet — a hauntingly beautiful subtropical wetland governed by a monsoonal climate. Only the Sud marshland of Sudan and the Pantanal of Paraguay and Brazil come close to duplicating this Floridian ecological treasure. Because we work at a state university adjacent to the Everglades, we have always felt privileged to have such a unique natural wonder, easily available for research, right at our doorstep. Although representing America's only Sud-like subtropical wetland, the Everglades has always frustrated geoscientists by obscuring the bulk of its true nature and history with surface layers of peat and muck.

The lack of scientific information about the geology of the Everglades has recently been reversed but at a great cost that has resulted in the devastation of the regional ecosystems. Over the past few decades, an explosion of housing, commercial, agricultural, and municipal development has rapidly encroached upon the edges of the Everglades, leaving only a tiny, dying remnant of the once immense wetland system. As a result of this development, an overwhelming amount of geological data has begun to pour out of these previously unexplored and inaccessible peripheral areas, most often from drainage canal digs, land and road fill quarries, and lake excavations for housing developments. We consider ourselves extremely fortunate to be able to conduct field research within the Everglades at exactly the right time in history. Because of the unrelenting frequency of new exposures due to construction, we have been forced to take on a "barbarian horde" research mentality, swooping into these elusive and ephemeral sites and gathering as much data as quickly as possible for posterity.

Our interpretation of all these new data is here combined with the viewpoints of many other types of Everglades geoscientists. These include hydrologists, structural geologists, geophysicists, sedimentologists, geomorphologists, stratigraphers, and paleontologists. We also throw our hats into the ring by bringing our own fields of expertise into the world of Everglades research: oceanography and malacology (the senior author) and remote sensing and geovisualization (the junior author). When the discoveries from all these geoscience specializations are synthesized into a single source, an entirely new picture of the regional geology appears. In this book, we hope to share, with all people who are fascinated by the Everglades, our new comprehensive view of the history of the great "River of Grass."

Edward J. Petuch
Charles E. Roberts
Boca Raton, Florida

Acknowledgments

Without the help and interest of the following individuals and organizations, this book would never have been produced. To all of them, we are greatly indebted. For allowing the senior author to conduct field research on their properties, we wish to thank: Enrique Tomeu and John Bates (Palm Beach Aggregates, Inc., Loxahatchee); M. Todd Lybrand, Jimmy Heimer, Elizabethann Phillips, and the Dickerson family (Dickerson Florida, Inc., Fort Drum Mine, Fort Drum, and the Indrio Mine, Fort Pierce); the Rucks family (Rucks Crystal Mine, Fort Drum); Mike Lorenz, Alfonso Alvarez, Patricia Rodriguez, Janet Barnwell, Andres Fanjul and the Fanjul family (Florida Crystals, Inc., Okeelanta Corporation Division, Everglades Agricultural Area); Hugh Cannon (Quality Aggregates, Inc., Sarasota); the Richardson family (APAC quarries, Sarasota); Robert Roach, Jose Dimas, Richard Rhodes, and Iris Shellhorn (Florida Rock Industries' Naples Quarry, East Naples); Ronald Capeletti and the Capeletti family (Capeletti Brothers quarries, Hialeah); Howard A. Griffin and the Griffin family (Griffin Brothers quarries, Fort Lauderdale); Allen Ridgdill (Ridgdill Quarry, Moore Haven); the late D.L. Brantley (Brantley quarry, Arcadia); the late Robin Weeks (Weeks Quarry, Lakeport); and Fred Smith (Smith Quarry, Okeechobee).

For the donation of valuable research specimens (many of them illustrated in this book) and for assistance in the field, we thank the following: Eddie Matchett, Clifford Swearingen, Jason Gabriel, Lannie Gustafson, Phyllis Diegel, Richard Duerr, Larry and Judy Haley, Dale Stream, the late Charles and Violet Hertweck, Meta Jones, Steve and Roxane Wilson, Donald Asher, Edward Volek, George Saloney, Jonathan Arline, Joanna Arline, Brian Schnirel, James Houbrick, Joseph Buchek, Herbert and Fonda Waldron, Richard Anderson, Michael Bruggeman, Harry Hyaduck, Gary Leonard, Harry Yingst, the late Vladimir Eismont, Susan Khan, Mary Mansfield, Richard and Diane Pennington, William and Cathy Dowling, Eric Kendrew, Marilyn Barkley, Anthony Cinelli, William Aley, Rudolph Pascucci, Josiah Strauss, Jack Spengler, Art Harmon, Stephen Bass, Frank Boer, Freddy St. Jude, and Theodore Davis. Special thanks go to Mardie D. Banks (Florida Atlantic University) for photographing all the specimens, setting up the digital files, and for helping to design the figures and to Clifford Swearingen for the pen-and-ink drawings on the title and dedication pages and at the end of the Introduction.

For reviewing, editing, and helpful comments, we thank Dr. Thomas Lodge (Thomas E. Lodge Ecological Advisors, Inc., Coral Gables), Dr. M.G. Harasewych (Smithsonian Institution), and Dr. Anton Oleinik (Florida Atlantic University). Thanks also to John Sulzycki, Patricia Roberson, and Gail Renard at Taylor and Francis for their invaluable advice and assistance in the production of this book.

Authors

Edward J. Petuch was born in Bethesda, Maryland, in 1949. Being raised in a Navy family, he spent many of his childhood years collecting living and fossil shells in such varied localities as California, Puerto Rico, Chesapeake Bay, and Wisconsin. His early interests in paleontology and marine biology eventually led to B.A. and M.S. degrees in zoology from the University of Wisconsin–Milwaukee. While in Wisconsin, his thesis work concentrated on the molluscan biogeography of coastal West Africa. There, he traveled extensively in Morocco, Western Sahara, Senegal, Gambia, Sierra Leone, and the Cameroons. Continuing his education, Petuch studied marine biogeography under Gilbert Voss at the Rosenstiel School of Marine and Atmospheric Science, University of Miami.

During that time, his dissertation work involved intensive collecting and working on shrimp boats in Colombia, Venezuela, Barbados, the Grenadines, and Brazil. After receiving his Ph.D. in oceanography in 1980, Petuch undertook 2 years of postdoctoral research on paleoecology with Geerat Vermeij at the University of Maryland. There, he also held a research associateship with the Department of Paleobiology, National Museum of Natural History, Smithsonian Institution, and conducted intensive field work on the Plio–Pleistocene fossil beds of Florida and the Miocene of Maryland.

Petuch has also collected fossil and living mollusks in Australia, Papua–New Guinea, the Fiji Islands, French Polynesia, Japan, the Mediterranean coasts of North Africa and Europe, the Bahamas, Mexico, Belize, Nicaragua, and Uruguay. This research has led to the publication of almost 100 papers. His eight previous books (*Atlas of Living Olive Shells of the World* (with Dennis Sargent), *New Caribbean Molluscan Faunas*, *Neogene History of Tropical American Mollusks*, *Field Guide to the Ecphoras*, *Edge of the Fossil Sea*, *Atlas of Florida Fossil Shells*, *Coastal Paleoceanography of Eastern North America* and *Cenozoic Seas: The View from Eastern North America*) are well-known reference texts within the malacological and paleontological communities. Presently, Petuch is a professor of geology in the Department of Geosciences, Florida Atlantic University, Boca Raton, Florida. He resides in Lake Worth, Florida, with his wife Linda and three children, Eric, Brian, and Jennifer. When not collecting or studying mollusks, Petuch leads an active career as a musician and member of the university-affiliated Cuvier Trio, playing recorders and the harpsichord and specializing in Baroque and Renaissance music.

Charles Roberts was born on the Gulf coastal plain at the edge of Houston in 1953. He grew up watching the decline of natural ecosystems as urban sprawl rolled over the prairies. He developed an interest in conservation, especially the reconstruction of past environments. In 1979 he received a scholarship at Vassar College, where his interest in geography and urban planning lead him to obtain a bachelor of arts degree in geography–anthropology. In 1984 he entered the master's degree program in geography at Pennsylvania State University, specializing in historical geography and cultural landscapes. For his Ph.D. degree he focused on the historical geography of regional transformations and remote sensing of metropolitan land-use transitions. He earned his Ph.D. in geography in 1992.

In 1990 he came to Florida Atlantic University, where he is currently an associate professor of geography in the Department of Geosciences. He has taught a remote sensing course sequence and a geovisualization course sequence as well as courses in urban geography, cultural landscapes, and human–environmental interactions. He has supervised 18 master's theses in geography, geology, and environmental science.

Table of Contents

Introduction — The Everglades: Final Landscape of the Okeechobean Sea

To anyone driving across extreme southern Florida from Fort Lauderdale and Miami to Naples, the first thing that overtakes you once you've left civilization is an overwhelming vision of the vast saw grass prairies of the Everglades. This immense subtropical flooded grassland also encompasses a myriad of small uninhabited islands, each covered with unique ecosystems containing cypresses, tropical hardwoods, and cabbage palmettos. As described by Marjory Stoneman Douglas in her *The Everglades: River of Grass* (1947), this singular North American ecological treasure has always fascinated, but also simultaneously repelled, European visitors and settlers. Even today, the very word "Everglades" conjures up a vision of steaming swamps inhabited by dangerous alligators,

View of the Everglades, along U.S. Highway 41 (Tamiami Trail) near Ochopee, Collier County, showing the typical saw grass prairies and small islands of cabbage palmettos. The latest Pleistocene–early Holocene Lake Flirt Formation (Flamingo marl facies) can be seen exposed along the edge of the canal.

Detail of a shell bed in the APAC quarry at Sarasota, Sarasota County, showing the rich diversity and abundance of fossil mollusks found in the Pliocene of the Everglades region. Some of the more obvious species include the giant Duplin Horse Conch (*Triplofusus duplinensis*, at center), the large scallop *Carolinapecten eboreus*, and the oyster *Hyotissa meridionalis*. These are all representative of the Pinecrest Member of the Tamiami Formation.

swarms of mosquitoes, and venomous snakes. Only the Native Americans found refuge in these marshlands, adapting to the harsh conditions and creating their own unique lifestyles.

Although classified as a subtropical wetlands area, the Everglades that we see today is actually the last remnant of a huge series of marine environments that stretches back over tens of millions of years. At first glance, no one would consider the Everglades to be part of the oceanic system, but simply scratch the surface to a few meters' depth and ancient seafloors will be exposed. These prehistoric marine worlds, one stacked on top of the other for thousands of meters, have created the topography that governs the water flow and sedimentary deposition within the Everglades. The spectacular subtropical world that was so poetically described by Marjory Stoneman Douglas is but a thin film covering the extensive record of Florida's ancient seas. Any geologic study of the Everglades region, consequently, must be approached from the oceanographic perspective.

Looking across the Everglades today, it is hard to imagine what its appearance would have been like when it was covered by a shallow tropical sea. This unique marine world, the Okeechobean Sea (named for Lake Okeechobee; see Petuch, 2004), was essentially landlocked for much of its later history, being separated from the Gulf of Mexico and Atlantic Ocean by a narrow semicircular chain of shallow oyster banks, coral reefs, and small coral islands. Imagine a cross between present-day Florida Bay, with its mangrove jungles and turtle grass beds, and the Bahamas, with their coral cays and reef complexes; this would have typified the world of the Okeechobean Sea during the past six million years. Throughout this time, the only connections with the open oceanic environment were through narrow cuts and straits between the reefs, island chains, and the Florida mainland. As sea levels rose and fell during the past 40 million years, the Okeechobean Sea also was under the influence of a large-scale rhythmic pattern, with alternating times of dry land followed by

This immense pile of mollusk shells, surmounted by the senior author, readily demonstrates the abundance of fossils available to researchers in the Everglades area. These mollusks are from the Fruitville Member of the Tamiami Formation and the photograph was taken in the Quality Aggregates pit #9, Sarasota, Sarasota County. Photo by Brian Schnirel.

refloodings. Over its history, the Okeechobean Sea basin was flooded extensively 12 different times, resulting in discrete marine worlds separated by hundreds of thousands or millions of years of dry, terrestrial environments.

The separate flooding intervals of the Okeechobean Sea — many possibly reflecting global warming events — are here referred to as *subseas* (Petuch, 2004). Each Okeechobean subsea produced its own unique set of marine environments and ecosystems. These, in turn, contributed to the stratigraphic units and geologic formations that were deposited within each subsea. Many of the geologic formations laid down in these marine systems contain some of the richest and best-preserved fossil faunas found anywhere on Earth. These beautiful fossils, particularly mollusk shells and corals, are powerful tools for determining both the age and boundaries of the Everglades' geologic formations. Utilizing the information contained in these fossil beds and their accompanying sedimentary environments, we here attempt to bring to life the worlds that are preserved below the Everglades region. Hopefully, we will give a new perspective on the historical geology of this American natural treasure, the final landscape of the Okeechobean Sea.

1 Fundamentals of Everglades Geology

The geology of the Everglades differs greatly from the other geologic frameworks of the adjacent Floridian Peninsula and the southeastern U.S. as it resulted from deposition within an enclosed tropical sea. Over the past hundred years, scientists have become aware of the difficulty in interpreting the geology of such a homogeneous carbonate area. This, coupled with the paucity of exposures within the harsh Everglades swampland, has made southern Florida the last heavily-populated area of the U.S. to be fully explored geologically. Our present understanding of the underpinnings of the Everglades results from the labors, sacrifices, and insights of some of the most dedicated scientists in the history of American geology. These rugged explorers, particularly the early workers, braved the heat, mosquitoes, and dangerous conditions to gather, incrementally, the information that we have assembled here.

HISTORY OF GEOLOGIC EXPLORATION IN THE EVERGLADES

Considering that southern Florida now houses one of the largest population centers in the U.S., it seems amazing that, just a little over a century ago, the area was geologically completely unexplored. In early 1886, Angelo Heilprin of the Philadelphia Academy of Natural Sciences undertook the first geologic exploration of the areas around Lake Okeechobee and the Caloosahatchee River. Accompanied by the sponsors of the expedition (Joseph Willcox and Charles H. Brock of Philadelphia), Captain Frank Strobhar, and the cook Moses Natteal, Heilprin investigated the " … mysterious body of water — Okeechobee … " on the schooner "Rambler." This first geologic research was published later in 1886 (first draft; second draft published in 1887) by the Wagner Free Institute of Science, and Heilprin expounded at length about his adventures in the *Okeechobee Wilderness*. Besides giving the first physiographic description of the northern Everglades region, Heilprin also described new species of living snakes and fishes and some of the more spectacular Everglades fossil shells. This pioneer work excited the scientific community and set the stage for a flurry of geologic and paleontologic research.

For more than a decade, all publications on the geology and paleontology of the Everglades stemmed from a single individual, William Healey Dall of the U.S. Geologic Survey and the Smithsonian Institution (honorary curator). After a preliminary report on the geology of Florida (1887), Dall initiated the *Contributions to the Tertiary Fauna of Florida*, a series of large monographic works that was published between 1890 and 1903 (also by the Wagner Free Institute of Science). Within this same series (1892), Dall also published the first geologic map of Florida, with the Everglades area being left conspicuously blank. Nothing of this scope has ever been attempted since, and the series stands out as one of the great masterpieces of American geology and paleontology. Dall continued publishing works on Floridian geology (a large monograph on the Tampa Oligocene and early Miocene agatized shells) until 1915. Without doubt, William Healey Dall can be considered the "Father of Everglades Geology."

Following Heilprin and Dall's pioneer works, the newly established Florida Geological Survey (formed 1907) immediately began publishing a series of annual reports on exploration of the regional geology. One of the most important of these was G.C. Matson and F.G. Clapp's 1909 preliminary stratigraphic scheme for the Floridian Peninsula (Annual Report 2). Taking into account the lack of natural exposures in the Everglades area, Matson and Clapp's work concentrated on the northern half of the state. In the same year, Samuel Sanford (1909) described and named the surficial limestones of coastal southeastern Florida and the Florida Keys. A decade later, E.H. Sellards (1919) published the first geologic section across the Everglades in the Florida Geological Survey Annual Report 12. As new drainage canals were being dug across the Everglades at that time, Sellards and his coworkers were able to retrieve the first data on what geologic formations existed below the surface peat and muck. All the new information accumulated by the Florida Geological Survey was ultimately synthesized by C.W. Cooke and S. Mossom (1929), who published the first comprehensive work on the geology of the entire state (Annual Report 20). Cooke and Mossom were also the first geologists to traverse the Tamiami Trail. In less than 30 years, the Everglades region had gone from *terra incognita* to being part of the known geologic continuum of the Floridian Peninsula.

As Dall had dominated the 1890s with his prolific works, Wendell C. Mansfield dominated the 1930s with a similar plethora of research and publications. Although his earlier works concentrated on the geology and paleontology of northern Florida, Mansfield began to investigate (in the early 1930s) fossils from newly exposed areas within the Everglades Basin. With his publications in 1931 and 1932 on new fossil mollusks and echinoderms, Mansfield was the first person to discover and name new paleospecies from the southern part of the Everglades. His last work (published posthumously in 1939), which dealt with the stratigraphy of the Everglades Basin, set the stage for all modern studies of the regional geologic framework. Mansfield and Dall, together, named hundreds of important new molluscan stratigraphic index fossils, and their names as the authors of species of fossil shells can be seen throughout this book. In addition to Mansfield, several other important workers described new Everglades molluscan index fossils in the 1930s, including Helen I. Tucker (later Tucker–Rowland), Druid Wilson (1932, 1933) and Maxwell Smith (1936). Most of the discoveries of Tucker and Wilson came from two sources: the construction of the Hoover Dike around Lake Okeechobee following the devastating 1928 hurricane and some of the first commercial quarries in southern Florida near Acline, Lee County.

During the span of World War II, there was a marked drop in the number of publications on Everglades geology. Besides a few scattered papers on new fossil mollusk species by Maxwell Smith and Thomas McGinty (in 1940, before our entrance into the war), only a single large work was produced that dealt with the regional geology. This impressive monograph, *The Geology of Florida* by C.W. Cooke (1945), was the culmination and synthesis of all previous research on the geology of the entire State. Although now dated in content, the book is still a valuable resource for the geology of southern Florida, giving detailed descriptions and illustrations of stratigraphic sections now long lost. Cooke arranged his book geochronologically and illustrated many key index fossils, and we have chosen to use his monograph as a model for this book. Along with Gerald Parker, C.W. Cooke also published the first major study of the groundwater and geologic framework of southern Florida (1944).

With the end of World War II, geologic research in the Everglades region resumed with an intensity not seen since the 1930s. Within a 5-year span, three classic works on the regional geology and paleontology were published: Axel Olsson and Anne Harbison's large monograph on the molluscan paleofauna of St. Petersburg (1953, accompanied by smaller works by Fargo and Pilsbry), Gerald G. Parker's tome on the hydrogeology of the Everglades area (1955; with detailed core logs and deep subsurface data), and Jules DuBar's detailed study on the stratigraphy of the Caloosahatchee River area (1958). These three works represented a quantum leap in our knowledge of the Everglades.

The 1960s saw a flurry of commercial development and road building within the Everglades area. Probably the most conspicuous project was the construction of the highway named Alligator Alley in Broward and Collier Counties in the early 1960s (now part of Interstate Highway 75, running from Fort Lauderdale to Naples). The preliminary canal dredgings and road-fill excavations produced the first deep transect across the central Everglades. This was a windfall for the regional geologists and paleontologists, allowing them to bulk sample stratigraphic units that had previously been inaccessible or known only from drill cores. New stratigraphic index fossils from "Alligator Alley" and around Lake Okeechobee were reported in a series of publications by Axel Olsson (1967) and Richard E. Petit (1964, 1968) and by Emily Vokes (in the *Tulane Studies in Geology* series; i.e., 1963, 1965, 1966). With the new data from canal excavations around southwestern Florida, Muriel Hunter (1968) was able to offer the first accurate interpretation of the late Miocene and Pliocene stratigraphy of the Everglades region. The same year (1968) also saw the first in-depth study of the geology and origin of the Florida Keys by J.E. Hoffmeister and H.G. Multer. This classic work complemented Hoffmeister and Multer's earlier study (with K.W. Stockman, 1967) on the oölitic limestones of Dade County and the Florida Keys (Miami Formation). Throughout the 1960s, the deepening and channelizing of the Kissimmee River was undertaken by the Army Corps of Engineers. The exposure of sediments in the spoil banks along the Kissimmee River allowed researchers to study, for the first time, the unusual and highly endemic fossil mollusks of that area.

The 1970s marked the beginning of major housing development and urban sprawl around the edges of the Everglades basin, and road building afforded accessibility to many new research sites, particularly along the Kissimmee River in the northern Everglades. At the same time, the demand for fill dirt (primarily for housing construction) produced the economic incentives for large quarries to be excavated all along the southwestern and southern Everglades area. Some of the most important of these included the Ashland Petroleum and Asphalt Corp. (APAC) quarry at Sarasota (originally the Warren Brothers and Richardson pit), the Mule Pen quarry at East Naples (later the Florida Rock Industries Naples quarry), and the quarries at Sunniland. Many important new index fossils were described from these sites, including new mollusks (S.C. Hollister, 1971; Emily Vokes, 1975, 1976) and corals (N.E. Weisbord, 1974). All the new data accumulated during the 1950s and 1960s allowed Jules DuBar to build upon his early work on the Caloosahatchee area and create the first workable stratigraphic scheme for the entire Neogene of the Everglades (1974). This work was particularly important because of its descriptions and clarifications of the then poorly known Pleistocene formations. Further research on the Pleistocene and Holocene was undertaken by Enos and Perkins (1977), who published the first detailed study of Quaternary depositional environments in the Everglades and Florida Bay.

The explosion of housing development that started in the 1970s, with increased numbers of landfill quarries and canal digs, escalated even more in southern Florida during the 1980s. Although a disaster for the ecosystems within the Everglades and the peripheral areas, the profusion of new excavations offered geologists many opportunities to study subsurface areas that were previously inaccessible. One of the more important discoveries that resulted from this land development was the uncovering of large, complex Pliocene coral reef systems along southwestern Florida. From data collected in canal digs and quarries in that area, J.F. Meeder was the first to describe the Pliocene reef limestones and associated carbonate facies found along the Collier County and Lee County coasts (1980, based upon his previous work published in a geology field trip guide in 1979). Subsequently, the senior author (E.J. Petuch) discovered a "mirror-image" system of Pliocene coral reefs under Dade County (the "Bird Road Dig" of Miami, 1986). When the data on the two contemporaneous reef tracts were later combined, it was shown that the east and west coast systems constituted a single, large contiguous coralline feature, the Everglades Pseudoatoll (Petuch, 1987).

Resulting from research in the APAC quarry at Sarasota, the senior author published the first reference stratigraphic column for the Pliocene of southern Florida earlier in the decade (Petuch, 1982; reproduced here in Chapter 4). The first attempt at bringing together geological, environmental, and anthropological studies into a single volume was undertaken by the Miami Geological Society in 1984, with the publication of "Environments of South Florida, Past and Present" (Patrick J. Gleason, editor). The hydrogeology of the east coast Pliocene reef tracts was studied extensively from 1985 to 1986, and the data were subsequently published by L.J. Swayze and W.L. Miller in 1987 (U.S. Geological Survey, Water Resources Division). Thomas M. Scott (Florida Geological Survey) published his large monographic work on the Hawthorn Group of Florida one year later. This pioneer work elucidated the underpinnings of the Everglades basin and set the stage for all subsequent research on the Miocene and early Pliocene of the Everglades area. Paleontological discoveries within the

Everglades area continued to be reported throughout the late 1980s, with works by Juan Parodiz (1988, on new mollusks) and the senior author (Petuch, 1987, 1988, 1989; new mollusks).

The 1990s saw a continuation of the intense paleontological research that started in the previous decade. Working at several new fossil sites, the senior author was able to describe over 350 new gastropods (Miocene to Pleistocene; Petuch, 1990, 1991, 1994, 1998; Oligocene, 1997). Similarly, several other workers added important new molluscan index fossils to the Everglades literature, including William Lyons (1991), Juan Parodiz and Jay Tripp (1997), and Geerat J. Vermeij and Emily Vokes (1997). The first integrated geochronological study of a southern Floridian Plio–Pleistocene section (APAC quarry, Sarasota) was published by Douglas Jones et al. in 1991, which complimented and supported William Lyons' reanalysis of the regional Plio–Pleistocene stratigraphy (1991). A collection of review papers, giving a broad overview of Floridian geology, was published as *The Geology of Florida* in 1997 (edited by Anthony Randazzo and Douglas Jones of the University of Florida). The volume contains contributions from several geologists and paleontologists who have made important advances in our knowledge of the Everglades region, including Walter Schmidt (Florida State Geologist), Thomas Scott (Assistant State Geologist), and Bruce J. MacFadden. Contributions and discoveries so far in the 21st century are reported on separately in the following chapters.

GEOMORPHOLOGY OF THE EVERGLADES REGION

The surficial geomorphology of the Everglades region controls the flow and impoundment of water throughout the entire wetland system. Within the scope of this book, we take into account all the main landforms south of a line running from Sarasota in the west to Vero Beach in the east. Besides the Everglades Physiographic Province itself, we also include contiguous areas that are either watersheds feeding into the main wetlands or peripheral areas that receive drainage from the Everglades. Scott (1997, Figure 5.1) includes most of this area within the Okeechobee Basin structural feature. The geomorphologic nomenclature used here is taken from William A. White's monumental work, *The Geomorphology of the Florida Peninsula* (1970). Our physiographic scheme, slightly modified from White, is shown here in Figure 1.1.

The Everglades region can be subdivided into three areas: an **Eastern Region**, including the Kissimmee River area, Lake Okeechobee, and the main Everglades saw grass prairies; a **Western Region**, including the complex Gulf of Mexico coastal areas, the Caloosahatchee River area, and the high pine forest areas of Lee and Collier Counties; and a **Southern Region**, including Florida Bay and the Florida Keys. The Everglades represents a true basin, with a low, flooded area surrounded by a rim of higher features. All three subdivisions have their own distinctive geologic frameworks and reflect bathymetric and depositional variations within the predecessor Okeechobean Sea.

Eastern Region

The eastern side of the Everglades is bounded in the north by the **Okeechobee Plain**, which extends all the way to Lake Kissimmee in Osceola County. This flat, swampy area contains the Kissimmee River and its meanders and ox-bow lakes (channelized by the Army Corps of Engineers in the early 1960s). The northward-tapering Okeechobee Plain is, in turn, bounded by the higher elevation areas of the **DeSoto Plain** to the west and the **Osceola Plain** (and its **Forty-Foot Ridge**) to the east. The Kissimmee River and several smaller rivers, such as Taylor Creek, are the principal water sources for Lake Okeechobee. This wide, shallow body of water, the second largest lake to be totally within the territory of the U.S., marks the northernmost edge of the Everglades Physiographic Province.

The actual Everglades, in the physiographic sense, is a wide, crescent-shaped saw grass prairie that extends from Lake Okeechobee and the Loxahatchee Slough southward to the mouth of the Shark Valley Slough at Florida Bay. Throughout its length, the Everglades is bounded on the east by the **Atlantic Coastal Ridge** and on the west by the **Immokalee Rise** (with its **Big Cypress Spur**). The orientation of these raised features causes the water flow to make a sharp bend to the

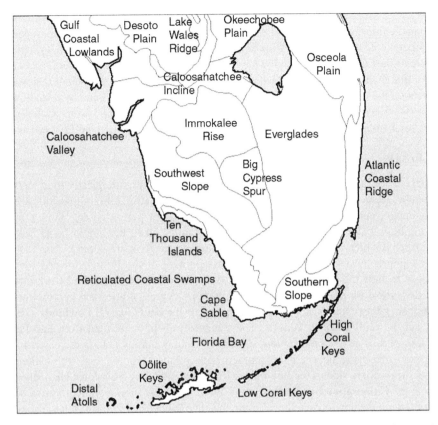

FIGURE 1.1 Geomorphologic map of southern Florida, showing the main surficial land forms of the Everglades area. (Adapted from White, 1970.)

west in the southern quarter section. The Palm Beach County section of the Everglades, south of Lake Okeechobee, has been developed for agriculture, primarily sugar cane, and is referred to as the Everglades Agricultural Area (EAA). Along the south, a chain of small, pine-covered hammocks, the **Pine Keys** (or "Everglades Keys") extend along the **Southern Slope** of the Atlantic Coastal Ridge and the **Southwestern Slope** of the Immokalee Rise. Although bisected by narrow, low marshlands (the **Transverse Glades**), the Atlantic Coastal Ridge, along with the Pine Keys, effectively landlocks the Everglades saw grass areas.

Within the Eastern Region, the Atlantic Coastal Ridge is of special interest. As originally conceived by Cooke (1945) and White (1970), this narrow, high ridge was thought to be composed of a linear series of transverse sand dunes (late Pleistocene–Holocene age) that run along the entire east coast of Florida, from Duval County in the north to Dade County in the south. Subsequent research (Petuch, 1986; Swayze and Miller, 1984) has shown that this feature is more complex than originally thought. The southern one third of the Atlantic Coastal Ridge, in Palm Beach, Broward, and Dade Counties, consists of a series of Pliocene zonated coral reefs (part of the Everglades Pseudoatoll; see Chapter 4), 10–30 m below the surface, which is covered by thin layers of late Pleistocene sands and oölitic limestone. The northern two thirds of the Atlantic Coastal Ridge is composed of the same transverse dunes, but they are built upon older sand features such as Plio–Pleistocene coastal lagoons and barrier islands. The surficial sand dunes produce the appearance of a single linear feature, whereas the deeper subsurface geology shows that the ridge is actually composed of two distinct structural features. The **Green Ridge** of Martin County, on the **Eastern Valley** lowlands east of the Forty-Foot Ridge, may represent a northern extension, along the paleocontinental shelf margin, of the Pliocene Pseudoatoll reef tracts.

For the Eastern Region, the Osceola and DeSoto Plains represented the southernmost edges of the Floridian Peninsula during Pliocene times. These two paleoshorelines were separated by a shallow embayment, the Kissimmee Embayment (see Chapter 3 and Chapter 4), which is still recognizable as the Okeechobee Plain lowlands. The Lake Okeechobee and Everglades depressions lie in a paleotrough feature (the Loxahatchee Trough; see Chapter 3 and Chapter 4) that was bounded by reef tracts on the east (now buried under the Atlantic Coastal Ridge) and by a shallow platform to the west (the Hendry Platform and Miccosukee Island; see Chapter 3 to Chapter 5). The flow of water in the Everglades still follows the contours of this paleotrough and Pliocene reef system.

WESTERN REGION

The western side of the area adjacent to the Everglades is geomorphologically more complex than the eastern side. Here, the dominant feature is the **Immokalee Rise**, a high limestone area covered by pine, palmetto, and cypress forests. Extending throughout parts of Lee, Hendry, and Collier Counties, the Immokalee Rise contains several small karstic lakes, the largest being Lake Trafford near Immokalee, Collier County. Adjacent to the Immokalee Rise are several areas that are intermediate in topography between the low saw grass prairies of the Everglades and the high pine forests. These include the **Big Cypress Spur** and the Fakahatchee Strand and Corkscrew Swamp of the **Southwestern Slope**, all of which are typified by cypress domes and cypress forests.

The northern edge of the Western Region extends into the **Gulf Coastal Lowlands**, **Gulf Coastal Lagoons**, and the **Gulf Barrier Islands**. These features all have resulted from late Pleistocene–Holocene transport of quartz sand by low-velocity longshore currents along the eastern side of the Gulf of Mexico. Running between the southern side of the DeSoto Plain (the **Caloosahatchee Incline**) and the northern side of the Immokalee Rise, and extending from Lake Hicpochee and Lake Okeechobee, the **Caloosahatchee Valley** is also one of the most prominent features of the Western Region. Containing the now-channelized Caloosahatchee River, the Caloosahatchee Valley separates the Gulf Coastal Lowlands from the Southwestern Slope of the Immokalee Rise and the dense mangrove forests of southern Collier and Monroe Counties. The mangroves, mostly red mangroves (*Rhizophora mangle*) and black mangroves (*Avicennia germinans*), form a myriad of small, densely packed islands (the Ten Thousand Islands of Collier and Monroe Counties) and impenetrable shoreline jungles which, together, form the **Reticulated Coastal Swamps** of northern Florida Bay.

Like the Atlantic Coastal Ridge, the Immokalee Rise is underlain by a Pliocene-aged coral reef tract. Extending from the Caloosahatchee Valley to the southernmost end of the Southwestern Slope, this reef system differs from its eastern counterpart in being much wider and better developed (see Chapter 4). East of the Immokalee reef systems, a wide carbonate platform (the Hendry Platform) extended to the western edge of the Everglades and the Loxahatchee Trough. This prominent Pliocene carbonate feature, in turn, was deposited on top of two large prograding deltas (the DeSoto and Immokalee Deltas; see Chapter 3). The Pliocene coral reef tract and underlying Miocene oyster banks (along the western edge of the Immokalee Rise and Southwestern Slope) funneled the clays and sands of these deltas into the Okeechobean Sea basin, filling the entire western two thirds of it. Farther north, the Gulf Coast Lowlands have recently been found to have been built upon an extensive Pliocene estuary (the Myakka Lagoon System; see Chapter 4). This estuarine system contained environments that were similar to those of the Kissimmee Embayment and Okeechobee Plain. In the extreme southwestern corner, **Cape Sable** stands out as a high area surrounded by mangrove lowlands. This island-like feature formed as an isolated coral bank during the Pliocene and later connected to the mainland.

SOUTHERN REGION

Geologically the youngest part of the Everglades area, the Southern Region is dominated by a single feature, the **Florida Keys**. Dating from the Sangamonian Pleistocene, the Keys archipelago consists of four main sections, each classified by its lithologic composition. These include the **High**

Coral Keys (running from Key Biscayne and Soldier Key to Plantation Key), the **Low Coral Keys** (running from Windley Key to Bahia Honda Key), the **Oölite Keys** (running from West Summerland and Big Pine Keys to Boca Grande Key, west of Key West), and the **Distal Atolls** (including the Marquesas Keys and the Dry Tortugas). Both the high and low coral keys are formed around cores of reefal limestone (the Key Largo Formation; see Chapter 7). The Oölite Keys represent an extension of the oölitic Miami Formation (see Chapter 7), which extends from the southern section of the Atlantic Coastal Ridge, across (and underlying) Florida Bay, to Boca Grande Key. The Distal Atolls are new, late Pleistocene–Holocene features, composed primarily of living coral reefs, cemented coral rubble, and mangrove peats (Marquesas Keys).

Florida Bay is, essentially, the last remnant of the Okeechobean Sea marine environment. This shallow feature, filled with small mangrove islands, carbonate mud banks covered with Turtle Grass (*Thalassia testudinum*), and coral and sponge bioherms, approximates what the Okeechobee Plain, Lake Okeechobee, and Everglades would have looked like during the Pliocene and Pleistocene. An impoverished mid-Pleistocene molluscan fauna still exists in the Florida Bay Turtle Grass beds (Petuch, 2004: 265), underscoring the oceanographic ties to the Okeechobean Sea.

GEOLOGIC FRAMEWORK OF THE EVERGLADES REGION

The upper 330 m of the Everglades region, the area covered in this book, encompasses 4 geologic groups, 18 formations, and 23 described members. Geochronologically, these units range from the Upper Eocene (Priabonian Age; Jackson Stage) to the Holocene. Over this 40 million-year time period, there were three main episodes of deposition, with each reflected in the resultant groups of formations. These include: the **Carbonate Platform Depositional Episode**, the **Upwelling-Deltaic Depositional Episode**, and the **Pseudoatoll Depositional Episode**. The geologic framework produced by these three depositional episodes is shown in Figure 1.2.

Based on the research outlined in Chapters 2 through 7 of this book, we here introduce, for the first time, seven new geologic members (the Cocoplum, Port Charlotte, Jupiter Waterway, Fordville, Fort Drum, Rucks Pit, and Okeelanta). We also resurrect and better define eight other members that were described earlier but have fallen into disuse. These include the Sarasota, Golden Gate, Fort Denaud, Bee Branch, Ayers Landing, Holey Land, Belle Glade, and Okaloacoochee. At the formational level, we here change the status of four units. The Bayshore and Murdock Station, originally described as members, are here elevated to formation status. The Buckingham and Fruitville, originally described as formations, are here demoted to member status. Our recommended nomenclatural scheme is put forth simply as a refinement of the stratigraphy that is currently recognized by the Florida Geological Survey and the U.S. Geological Survey. Because of the pending status of many of these units, our stratigraphic scheme should only be considered to be informally proposed at the present time.

THE CARBONATE PLATFORM DEPOSITIONAL EPISODE

This period, at roughly 40 million years B.P., represents the incipiency of the deep structural features found below the Everglades area. At that time, Florida consisted of a shallow carbonate bank (the Ocala Bank) off the coast of southern Georgia and resembled a smaller version of the Recent Bahamas Banks (more detail given in Chapter 2). The southern edge of the Ocala Bank was in the vicinity of modern Orlando, Orange County, and the northern edge was near Gainesville, Alachua County. To the south, the bank sloped gently to the Straits of Florida, with the mean water depth in the Everglades region being around 200–300 m. Because the Ocala Bank was so far off the coast of Georgia, and separated by the wide Suwannee Strait, very little, if any, continental siliciclastic material extended to the southern slope. Here, instead, a constant rain of carbonate fines poured off the bank and produced a wide, thick sedimentary draping. By the late Eocene (Jackson Stage), these carbonates had accumulated to produce the **Ocala Group** of formations, including the basal **Inglis**

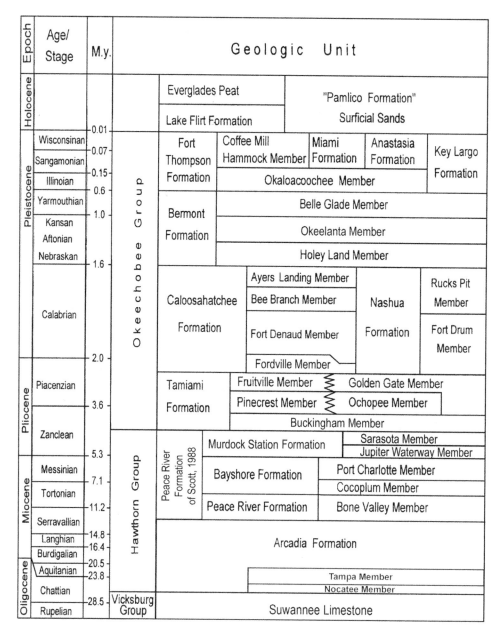

FIGURE 1.2 Generalized stratigraphic column for the upper 300 m of the Everglades and adjacent areas. The eight formations of the carbonate-rich Okeechobee Group (Tamiami, Caloosahatchee, Nashua, Bermont, Fort Thompson, Miami, Anastasia, and Key Largo) were all deposited during a time span ranging from the early Pliocene to the latest Pleistocene. The four formations of the phosphate-rich Hawthorn Group (Arcadia, Peace River, Bayshore, and Murdock Station) were deposited during a time span ranging from the late Oligocene to the early Pliocene. The single formation belonging to the Vicksburg Group (Suwannee) was deposited during the early-to-late Oligocene, whereas the three formations of the Ocala Group (Inglis, Williston, and Crystal River) (not shown here) were deposited during the late Eocene. The Ocala Group is missing under much of the Everglades.

Formation, the middle **Williston Formation**, and the upper **Crystal River Formation**. These were deposited on top of the older Eocene (Claiborne Stage) **Avon Park Formation**.

At the end of the Eocene, an incredible event took place along the mid-Atlantic coastline of the eastern U.S. — one that would permanently alter the shape of the Floridian Peninsula to the south. As the Yucatan Peninsula had been hit by an enormous asteroid 65 million years ago, ending the Cretaceous Period, the Chesapeake Bay area was hit by another, slightly smaller asteroid 36 million years ago (Poag, 1999; Poag et al., 2004). This asteroid may have originally been a much larger body, but it broke into innumerable smaller fragments prior to impact. The largest of the fragments struck both Popigai, Siberia, and the area of the present-day mouth of Chesapeake Bay. The ejecta- and tektite-strewn field of the Chesapeake impact passed directly over the Ocala Bank and extended all the way to Argentina. The resultant crater was over 80 km in diameter and the impact produced a series of megatsunamis over 300 m high.

Refracting down the coast, these megatsunamis, virtually undiminished in energy, would have passed right over the Ocala Bank, scouring off all unconsolidated sediments. The main platform and island chains, being already consolidated, would have remained intact, as would have the underlying Avon Park limestones. All the unconsolidated fines of the Ocala Group on the seafloor of the southern slope (present-day Everglades area), however, would have been removed, leaving the Avon Park as the "new" posttsunami seafloor. This unconformity is still present under parts of the Everglades and is discussed in more detail in Chapter 2. Extreme global cooling ("nuclear winter") took place after the impact, lowering sea levels and exposing southern Florida to subaerial conditions. During this regressive time, the southeasternmost edge of the Florida Platform (present southern Everglades) underwent heavy dissolution by groundwater and aquifers, and formed a low depressional feature (see Chapter 2).

The mid-Oligocene (late Rupelian Age) saw a return to high tropical conditions on the Floridian Peninsula, and rising sea levels flooded much of the old Eocene seafloor areas. This tropical inundation led to renewed carbonate deposition, producing the **Suwannee Limestone**. This last carbonate platform is here placed within the **Vicksburg Group** of the Mississippi Embayment, primarily because of its lithologic resemblance to the Flint River Formation of Georgia and the Salt Mountain Formation of Alabama. All three of these formations share many index fossil species with the Byram Formation of Mississippi (a classic Vicksburg Group unit). As is seen in Chapter 2, the Suwannee Limestone is the first continuous, uninterrupted carbonate surface within the Everglades area since the mid-Eocene Avon Park Formation. By the late Oligocene (early Chattian Age), sea levels again regressed, and high tropical carbonate deposition ceased in southern Florida. This marks the end of the Carbonate Platform Episode.

THE UPWELLING-DELTAIC DEPOSITIONAL EPISODE

From the latest Oligocene (late Chattian Age) to the early Pliocene (Zanclean Age), deposition within the Everglades area changed dramatically. This resulted from two main factors: the establishment of permanent large-scale upwelling systems off western Florida and the accompanying high-productivity marine conditions and plankton blooms (Petuch, 2004: 139), and the southward progradation of giant deltas off the DeSoto Plain highlands region (see Chapter 3). Instead of the Bahamas-like carbonate environments of the Eocene to late Oligocene, the Okeechobean Sea now supported continental terriginous-type deposition, producing thick beds of clay minerals, quartz sand, and phosphorites. The deltas contributed the clay and sand component, whereas the offshore upwelling systems contributed cooler water conditions and abundant phosphate-rich plankton resources. The result was the deposition of the **Hawthorn Group** (Scott, 1988).

As presently understood, the Hawthorn Group of the Everglades area comprises four distinct formations: the **Arcadia Formation** (and its **Nocatee** and **Tampa Members**), which extends from the latest Oligocene (late Chattian Age) to the early Serravallian Age of the Miocene; the **Peace River Formation** (and its **Bone Valley Member**), which encompasses the late Serravallian Age and early

Tortonian Age of the Miocene; the **Bayshore Formation** (and its **Cocoplum** and **Port Charlotte Members**), which encompasses the late Tortonian and Messinian ages of the Miocene; and the **Murdock Station Formation** (and its **Jupiter Waterway** and **Sarasota Members**), which was deposited during the early Zanclean Age of the Pliocene. The clay- and phosphorite-rich Hawthorn Group is bounded by transitional facies containing higher percentages of carbonates. This is particularly noticeable in the limestone of the Tampa Member of the Arcadia Formation (basal Hawthorn Group) and the rich carbonate bioclasts of the Sarasota Member of the Murdock Station Formation (uppermost Hawthorn Group). In total, Hawthorn Group deposition lasted for over 20 million years.

THE PSEUDOATOLL DEPOSITIONAL EPISODE

Although confined to only a 4 million-year time period, this depositional episode produced the structural predecessor of the modern Everglades. In the late Zanclean Pliocene, the upwelling systems of western Florida began to diminish and phosphorite deposition ceased. The subsequent warmer water conditions allowed for the return of carbonate deposition on a large scale. By this time, the deltaic facies of the Hawthorn Group had filled the western two thirds of the Okeechobean Sea, leaving only a shallow bank. Concurrently, oyster banks that grew along the margins of the Okeechobean Sea during the Miocene had now built up to the sea surface. These two types of features, combined, produced the underpinnings of the Everglades Pseudoatoll (see Chapter 4).

The return of high tropical water conditions in southern Florida allowed for the growth of the largest and most species-rich coral reefs ever seen in the Cenozoic of the eastern U.S. Within the coral reef arcs of the Everglades Pseudoatoll (which grew upon the older Hawthorn oyster bars), a high diversity of depositional environments was produced, including shallow carbonate banks, patch reef bioherms, Turtle Grass beds, and mangrove mud flat estuaries. These environments, which persisted from the late Zanclean Pliocene to the Wisconsinan Stage of the Pleistocene, produced the sediments that characterize the **Okeechobee Group**. This group of eight formations is unique to the Everglades region and southern Florida (the Okeechobean Sea area) and differs from the underlying Hawthorn Group in being essentially devoid of phosphorites, in having varying amounts of quartz sand, and in having a much higher percentage of carbonates. The concept of unifying all the carbonate-rich Plio–Pleistocene formations into a single unit was originally informally proposed by Scott (1992) as the "Okeechobee Formation," with the traditional Everglades formations being subordinated as members or faunizones. Subsequent research has shown that the geology of the area is too complicated and that Scott's Okeechobee Formation does not fully describe the depositional patterns and cyclostratigraphy of the area. Taking this into account, the senior author (Petuch, 2004: 18) elevated the Okeechobee Formation to group status, allowing for the retention of the traditional formational nomenclature and for the recognition of the lithologic uniqueness of the Everglades stratigraphic units.

The incipiency of the Everglades Pseudoatoll is seen in the depositional patterns of the **Tamiami Formation** (and its **Buckingham**, **Pinecrest**, **Ochopee**, **Fruitville**, and **Golden Gate Members**), extending from the late Zanclean to late Piacenzian Pliocene. During this time, massive zonated coral reefs encircled the Okeechobean Sea, thick layers of carbonate fines buried the old deltas, and organic-rich sand deposits were put down in tropical estuaries at the mouths of rivers along the northern edge of the pseudoatoll lagoon. The facies produced by this diversity of environments are discussed in Chapter 4. The patterns of deposition seen during Tamiami time continued on into the Calabrian Age of the early Pleistocene, resulting in the deposition of the **Caloosahatchee Formation** (with its **Fordville**, **Fort Denaud**, **Bee Branch**, and **Ayers Landing Members**). An extension of the eastern coastal lagoon environments also occurred at this time, resulting in the deposition of the **Nashua Formation** (and its **Fort Drum** and **Rucks Pit Members**). The Caloosahatchee and Nashua depositions are discussed in Chapter 5.

The onset of major and frequent glaciations and sea level drops during the middle and late Pleistocene accelerated the demise of the Everglades Pseudoatoll and ushered in the formation of the modern Everglades. During the sea level drops of glacial stages, the Okeechobean Sea was completely emergent and the Everglades basin contained a series of large freshwater lakes. During warmer interglacial times, the Okeechobean Sea would again reflood, filling in depressional features with biogenic carbonates and erosional fines, and causing the enlargement of older island chains by transported beach sands and mangrove forests. The sequential infilling of the Everglades Pseudoatoll, including emergent times with freshwater lakes and peripheral brackish lagoons, is evident in the depositional patterns of the **Bermont Formation** (with its **Holey Land**, **Okeelanta**, and **Belle Glade Members**) and the **Fort Thompson Formation** (with its **Okaloacoochee** and **Coffee Mill Hammock Members**) (see Chapter 6 and Chapter 7). Late Pleistocene beach deposits, coral reefs, and reef-associated oölite banks produced the uppermost formations of the Okeechobee Group: the **Anastasia**, **Key Largo**, and **Miami Formations**, respectively. This last part of the Pseudoatoll Depositional Episode is discussed in Chapter 7.

By the Holocene, the shallow slough-like structure of the Everglades was established. In light of present knowledge, the water flow within the Everglades system can now be seen to be guided by geomorphological features that were established as far back as the late Miocene. The formations of the Hawthorn and Okeechobee Groups, and their accompanying structures, are covered by a thin veneer of Recent sediments. These include three main types: peats and soils (primarily the **Everglades Peat**), sands (primarily wind-blown dunes and aeolianites of the **Pamlico Formation**), and freshwater calcilutites (the lime muds and "snail marls" of the **Lake Flirt Formation**). These three types of surficial sediments interfinger across the Everglades area, depending upon the subsurface topography. Only Recent Florida Bay and the Florida Keys retain an Okeechobean Sea environment, with coral reefs, extensive carbonate banks covered with Turtle Grass, and mangrove forests. All of these Holocene sediment types are discussed in Chapter 7.

EVERGLADES MOLLUSK AGES AND RELATIVE DATING

Anyone conducting geological research in the Everglades area, either in quarries, in canal excavations, or by using drillers' cores, becomes immediately aware of the abundance of molluscan shells that appear in almost every exposure or sample. Any geologic unit above the late Miocene is literally packed with beautifully preserved specimens, frequently exhibiting an amazingly rich species diversity. These fossils often represent the actual shell specimen, little changed by diagenesis, and occasionally retain the original color pattern. Some fossil shells, such as those of the Suwannee Limestone or the Tampa Member of the Arcadia Formation, are siliceous pseudomorphs, being completely replaced by multicolored agate (such as the Florida State Rock, the Tampa agatized corals). Other types of shell assemblages, such as those of the upper Hawthorn Group, are skewed by selective dissolution, with all the softer aragonite being dissolved away (often present only as molds) and with only the harder calcitic specimens surviving.

Regardless of preservation, if a stratigraphic unit contains any type of molluscan fossil, it can be dated to a relatively narrow time frame. This relative dating technique was originally applied to Everglades geochronology as far back as Dall (1890–1892), who used molluscan index fossils to date the Caloosahatchee Formation as being of Pliocene age. It was later shown that only the lower beds dated from the Pliocene and that the upper members were actually earliest Pleistocene. Considering the time in which Dall worked, however, this was an amazingly accurate determination. Mansfield (1937), using the same technique, correctly demonstrated that the Suwannee Limestone was of Oligocene age and that the Tampa Member of the Arcadia Formation dated from the latest Oligocene (later, also found to extend into the Aquitanian Miocene). Subsequent workers, such as Hunter (1968) and the senior author (Petuch, 1982; 1993), have since refined the dating accuracy by biostratigraphic correlations with other, better-studied faunas along the Atlantic Coastal Plain and northwestern Florida.

The patterns of marine molluscan evolution and extinction within the Everglades region closely approximate those used to define the North American Land Mammal Ages. This close correlation shows that the same ecological and climatological catastrophes that affected terrestrial environments also had extremely negative impacts on the contemporaneous marine environments. As most of the fossil beds in the Everglades region were deposited in marine conditions, terrestrial mammal ages are often of little practical use in local relative dating studies. To address this problem, the senior author devised a scheme involving three mollusk ages for the Plio–Pleistocene beds of the Everglades area (Petuch, 1993; 1994). This was based on molluscan index genera, each with sharply demarcated time ranges. We here have expanded and refined this original scheme to include all the molluscan faunas from the Oligocene to the Holocene. We now recognize five mollusk ages within the past 35 million years of Everglades history. These are the **Hernandoan Mollusk Age**, the **Charlottean Mollusk Age**, the **Kissimmean Mollusk Age**, the **Bellegladean Mollusk Age**, and the **Lakeworthian Mollusk Age**. Hopefully, workers in the local geosciences will find our dating scheme useful for determining both the times of deposition and formational boundaries. Our Mollusk Ages, then, can be considered proxies for the Land Mammal Ages.

THE HERNANDOAN MOLLUSK AGE

Named for Hernando County, site of many classic Suwannee and Tampa molluscan collecting localities (Mansfield, 1937), this mollusk age is exactly correlative with the Whitneyan and Arikareean Land Mammal Ages (35–20 million years). The distinctive faunas of this time period include those of the Suwannee Limestone and the Tampa Member of the Arcadia Formation. Examples of many of these are shown in Chapter 2 and some of the principal index genera shared by both formations include:

Gastropoda
Calvertitella
Orthaulax
Amauropsis
Ampullinopsis
Pterynotus (not found in Florida fossil beds after Tampa time)
Spinifulgur
Omogymna
Falsilyria

Although sharing these taxa with the Tampa Member, the Suwannee fauna also contains a large number of distinctive endemic genera. Some of these unique Suwannee taxa include:

Gastropoda
Cestumcerithium
Prismacerithium
Telescopium (North American species complex)
Suwannescapha

The Tampa Member, likewise, contains a large number of endemic genera. Many of these are also known from the late Oligocene–early Miocene formations in northern Florida and the Carolinas, but their presence in the Tampa fauna is the only known occurrence within the Everglades region. Some of these include:

Gastropoda
Floradusta
Loxacypraea
Doxander (only known record in the Americas)
Ecphorosycon
Tritonopsis

Because of groundwater infiltration and leaching during the deltaic episode of Everglades deposition, the shell beds of the subsequent Arcadia and Peace River Formations were completely dissolved. As a result, there exists a large gap in the Everglades mollusk age timeline, corresponding to the Hemingfordian and Barstovian Land Mammal Ages (20 to 12 million years). No mollusk age is given for this large hiatus.

THE CHARLOTTEAN MOLLUSK AGE

Named for Charlotte County, site of the best known and best preserved late Miocene Everglades shell beds (Hunter, 1968), this mollusk age is exactly correlative with both the Clarendonian and Hemphillian Land Mammal Ages (12 to 4.5 million years). The faunas of the Charlottean Age are found in the heavily leached formations of the upper Hawthorn Group (the Bayshore and Murdock Station Formations). Because of this selective dissolution, only calcite shells are present, giving a diminished view of the original molluscan fauna. The surviving fossils, however, are large and abundant and give a distinctive look to the upper Hawthorn beds. Examples of many of these are shown in Chapter 3, and some of the principal index genera include:

Gastropoda
Calusathais
Globecphora
Ecphora gardnerae species complex (thick-ribbed species)
Zulloia
Bivalvia
Gigantostrea
Ostrea compressirostra species complex
Chesapecten palmyrensis-septenarius species complex

Although known from a few rare specimens found in the younger Tamiami Formation, the scallop genus *Chesapecten* (large species in the *Chesapecten middlesexensis-jeffersonius* species complex) is especially common in, and is indicative of, Charlottean-aged fossil beds.

THE KISSIMMEAN MOLLUSK AGE

Named for the Kissimmee River excavations (during the channelization of the early 1960s) which exposed some of the richest late Pliocene fossil beds ever seen in Florida, this mollusk age is exactly correlative with the entire Blancan and earliest part of the Irvingtonian Land Mammal Ages (4.5 to 1.6 million years). The Kissimmean fauna existed during the time of the maximum development of the Everglades Pseudoatoll. Because of the high tropical conditions and great diversity of habitats available within the pseudoatoll area, the molluscan faunas underwent species radiations that were unprecedented in the fossil record of the southeastern U.S. The faunas of this time period include those of the Tamiami, Caloosahatchee, and Nashua Formations and contain some of the first fossil mollusks collected from the Everglades region (Heilprin, 1886). Examples of many of these are found in Chapters 4 and 5, and some of the principal index genera that are endemic to the Everglades region include:

Gastropoda
Pahayokea (subgenus of *Siphocypraea*)
Siphocypraea
Acantholabia
Echinofulgur
Calophos (Floridian species complex)

Cymatophos (Floridian species complex)
Hystrivasum
Mansfieldella
Toroliva

Although sharing many of these taxa with the Caloosahatchee Formation, the Tamiami fauna also contains a large number of distinctive endemic genera that can be used as index fossils. Of special interest is an evolutionary explosion of large bernayine cowries with 37 described species in seven generic groups. This radiation of cowries is present in almost every facies of the Tamiami Formation, often with great numbers of individuals. Some of the endemic Tamiami taxa include:

Gastropoda
Calusacypraea
Myakkacypraea (subgenus of *Calusacypraea*)
Pseudadusta
Tropochasca
Lindafulgur (still living in deep water in the Gulf of Mexico)

Three other very important Tamiami index fossil groups are shared with contemporaneous Piacenzian Pliocene formations in the Carolinas and Virginia, and these include:

Gastropoda
Eichwaldiella
Akleistostoma
Latecphora

Although known from only two species in the Duplin and Yorktown Formations of Virginia and the Carolinas, the cowrie genus *Akleistostoma* radiated into at least 13 endemic Floridian species during Tamiami time. In late Kissimmean time, at the end of the Blancan Land Mammal Age (the Tamiami–Caloosahatchee boundary), the great Kissimmean cowrie radiation suffered a major extinction, with the genera *Akleistostoma*, *Calusacypraea*, and *Pseudadusta*, and the subgenus *Myakkacypraea*, disappearing from the fossil record. Only the sympatric genus *Siphocypraea* and subgenus *Pahayokea* survived into Caloosahatchee time. This extinction is a good marker for the Tamiami–Caloosahatchee formational boundary. The subgenus *Okeechobea*, originally thought to be confined to the Caloosahatchee Formation, may actually be present in the Tamiami Formation as a cluster of still-undescribed species from the Kissimmee Valley and Okeechobee Plain areas.

The Kissimmean faunas of the Everglades region also shared a large number of chronologically restricted species with the late Pliocene and early Pleistocene formations of the Carolinas and Virginia. These index taxa all disappear at the same time (early Pleistocene), demonstrating that the same environmental degradation and extinction-producing events occurred along the entire southeastern U.S. Some of these widespread genera include:

Gastropoda
Trossulasalpinx
Terebraspira
Brachysycon
Globinassa
Paranassa
Scalanassa
Contraconus
Cymatosyrinx

Bivalvia
Cunearca
Dallarca
Rasia
Stralopecten

As in the cowrie genus *Akleistostoma*, the gastropod genera *Trossulasalpinx*, *Terebraspira*, and *Contraconus* were represented in the Carolinas and Virginia by only a few species, whereas they underwent extensive species radiations in Florida. The Everglades Pseudoatoll region, then, represented the center of speciation for most of these characteristic Kissimmean taxa. Their presence in Virginia and the Carolinas actually constituted the northernmost edge of their ranges.

THE BELLEGLADEAN MOLLUSK AGE

Named for the city of Belle Glade, Palm Beach County, site of the first quarries where the typical fauna was collected (Hoerle, 1970; McGinty, 1970), this mollusk age correlates with the last three quarters of the Irvingtonian Land Mammal Age (1.6 million–150,000 years). The distinctive faunas of this time period include those of the Bermont Formation and its three members. The Bellegladean fauna is an interesting transition between the previous rich Kissimmean faunas and the subsequent impoverished faunas of the latest Pleistocene and Holocene. Typically, the Bellegladean fauna contains both relictual Kissimmean taxa and newly evolved endemic genera. Examples of many of these are shown in Chapter 6, and some of the more important index genera include:

Gastropoda
Pseudozonaria (Florida species complex)
Lindoliva
Eurypyrene
Seminolina (Lake Okeelanta species complexes)

While the coastal waters of the southeastern U.S. and the Gulf of Mexico became cold and nontropical during much of Bellegladean time, the interior of the nearly landlocked pseudoatoll stayed much warmer and subtropical. This tiny remnant of previous tropical times acted as a refugium, housing the last vestiges of the older Kissimmean fauna. Found only in the Okeechobean Sea during the mid-Pleistocene were the following:

Gastropoda
Cerithioclava
Jenneria
Malea
Oliva (Porphyria)
Calusaconus
Bivalvia
Carolinapecten
Conradostrea
Armamiltha
Miltha
Semele perlamellosa species complex

Although containing many relictual Kissimmean taxa, the Bellegladean faunas can always be differentiated from those of the underlying Caloosahatchee Formation by lacking large Kissimmean genera such as *Contraconus*, *Siphocypraea*, *Terebraspira*, and *Hystrivasum*. By the Illinoian Glacial

Stage, all the Kissimmean relicts and Bermont endemics were extinct. Only a highly impoverished molluscan fauna continued on into the latest Pleistocene.

THE LAKEWORTHIAN MOLLUSK AGE

Named for the city of Lake Worth, Palm Beach County, site of the first good exposures from this time (during housing construction), the Lakeworthian Mollusk Age is correlative with the early part of the Rancholabrean Land Mammal Age (150,000 to 75,000 years). The faunas of this time are contained primarily within the Fort Thompson Formation, although broken specimens of mollusks are sometimes seen in the Anastasia Formation coquinas, the Miami Formation oölites, or the Key Largo Formation coral limestones. Having essentially a modern southern Floridian molluscan assemblage, the Lakeworthian faunas also contain a few relictual taxa and several extant species that no longer occur in eastern Florida. Because of the inclusion of a few anomalous taxa, these latest Pleistocene faunas can always be told apart from assemblages found in Holocene or Subrecent deposits. Examples of typical Lakeworthian genera and species are shown in Chapter 7, and some of the more important include:

Gastropoda
Pyrazisinus
Cerithidea beattyi species complex (now confined to the Bahamas and Greater Antilles)
Strombus lindae species complex (oldest Lakeworthian only)
Sinistrofulgur perversum species complex (now confined to the western and southern Gulf of Mexico)
Vokesinotus perrugatus (now confined to western Florida)
Solenosteira cancellarius (now confined to western Florida and the northern Gulf of Mexico)
Turbinella (oldest Lakeworthian only, extinct in Recent Florida)

During the Wisconsinan Glacial Stage, the potamidid genus *Pyrazisinus* became extinct, having almost survived to the Holocene. This unique Everglades animal, with the oldest known species being found in the Suwannee Limestone, spent its entire evolutionary history within the confines of the Okeechobean Sea. Over this 35 million-year span, *Pyrazisinus* lived only in the extensive mangrove forests and estuarine mud flats that ringed the Everglades Pseudoatoll and evolved into at least 20 species. The sea level drops during the early Wisconsinan Stage would have drained the mangrove habitats of *Pyrazisinus*, and these, coupled with colder air and water temperatures, destroyed the last dwindling populations in the Loxahatchee Trough area.

PALEOCEANOGRAPHY OF THE EVERGLADES

Throughout this chapter, we have referred to the Okeechobean Sea primarily in the broad sense, as a single body of water covering the Everglades area. Over the past 40 million years, the Okeechobean basin actually has been flooded by 12 separate major marine incursions. Each of these, in turn, had intervals of minor sea level fluctuations superimposed upon the larger transgressive pattern. As will be seen in the following chapters, the larger transgressions resulted in the deposition of the Everglades geologic formations whereas the minor transgressions resulted in the deposition of their members and beds. The intervening regressive intervals varied in time from millions of years (between the large transgressive intervals) to thousands of years (between the smaller transgressive intervals). About one third of the 40 million-year history of the Okeechobean Sea is lost due to multiple extended periods of terrestrial conditions.

In an attempt to visualize what Florida may have looked like during the major Okeechobean transgressions, we pooled our academic specialties of paleoceanography and geovisualization to produce a new geosciences application: simulated space shuttle imagery. In the first step of our new application, the paleoenvironments and paleoceanography of the Everglades area were determined for each time interval. The senior author contributed to this part of the synthesis by studying and interpreting fossil shell beds throughout the Everglades area. With over 30 years of research in over 100 quarries, canal digs, well core sites, and construction excavations, the senior author was able to determine, with a high degree of accuracy, the age, bathymetry, and marine environments of large areas of the Okeechobean Sea. In the second step, the junior author took the paleoceanographic data, along with the regional geochronology, and superimposed this upon a space shuttle (hand-held) photograph of the Floridian Peninsula from an altitude of 100 mi. This gave us a sequence of preliminary illustrations, each containing the approximate geomorphology and bathymetry of the Everglades region at different time intervals.

The third step in our simulated space shuttle image methodology incorporated satellite image interpretation by the junior author and benthic ecology interpretation by the senior author. As a palette for image cloning, we used a space shuttle image of Andros Island, Bahamas, from the same altitude. Knowing the present-day marine environments of that area, we identified and selected classic tropical benthic and emergent environments as they appear from 100 mi in space. These included Turtle Grass (*Thalassia*) beds, coral reefs, carbonate banks, deep carbonate lagoons, trough features, mangrove jungles, and estuaries — all environments that dominated the Okeechobean Sea. A fragment of each of these modern-day Bahamian environments was then transferred to our rough paleomaps, with each fragment being placed on its corresponding paleoenvironment. Using a modified Adobe Photoshop program, the junior author cloned the individual environmental images and filled in the intervening blank areas. This image completion was guided by surface topography maps, stratigraphic data, and previous subsurface geomorphological studies.

The end result of our new geovisualization application is illustrated here, in a series of color simulated space shuttle images showing the probable appearance of the Okeechobean Sea at different times over 35 million years. We have also taken these subsea images and combined them with images of emergent times (marine regressions), producing an animation showing the growth of the Floridian Peninsula over 40 million years. This animation, with timeline annotations and paleogeographic features, is contained in the DVD appended to the back of this book.

As discussed earlier in this chapter, the depositional and chronological subdivisions of the Okeechobean Sea are here referred to as "subseas." Between the time of the Ocala Sea and the megatsunamis (see Chapter 2) and Recent Florida, we recognize 11 distinct subseas of the Okeechobean Sea. These oceanographic subdivisions were named and described by the senior author (Petuch, 2004) and include (from oldest to youngest): the **Dade Subsea** (named for Dade County), the **Tampa Subsea** (named for Tampa, Hillsborough County, and the Tampa Member of the Arcadia Formation), the **Arcadia Subsea** (named for Arcadia, DeSoto County, and the Arcadia Formation), the **Polk Subsea** (named for Polk County), the **Charlotte Subsea** (named for Charlotte County), the **Murdock Subsea** (named for Murdock Station, Charlotte County, and the Murdock Station Formation), the **Tamiami Subsea** (named for the Tamiami Trail in Dade and Collier Counties, and the Tamiami Formation), the **Caloosahatchee Subsea** (named for the Caloosahatchee River and the Caloosahatchee Formation), the **Loxahatchee Subsea** (named for Loxahatchee, Palm Beach County), the **Belle Glade Subsea** (named for Belle Glade, Palm Beach County), and the **Lake Worth Subsea** (named for Lake Worth, Palm Beach County). These subsea intervals are shown in the color simulated space shuttle images which immediately follow page 8:

Figure 1.3 — The **Dade Subsea** (Rupelian and early Chattian Oligocene). Formed after the massive catastrophism of the late Eocene megatsunamis and meteor impacts (see Chapter 2), the Dade Subsea area contained the first geomorphological configuration that can be considered the predecessor of the Recent Floridian Peninsula. At this time, Florida consisted of a large, elongated island (Orange Island) that was separated from the Georgia mainland by a wide, shallow channel (the Suwannee Strait). As can be seen in this illustration, both the Suwannee Strait and the west coast of Orange Island contained extensive coral reef complexes. South of Orange Island, two small arcs of deep carbonate banks mark the edge of what will be, 30 million years later, the Everglades. Orange Island still exists today as the Ocala Arch and highlands area. This subsea represented the last phase of the Carbonate Platform Depositional Episode.

Figure 1.4 — The **Tampa Subsea** (late Chattian Oligocene to early Aquitanian Miocene). The template set down by the Dade Subsea was enlarged and expanded within the Tampa Subsea area (see Chapter 2). At this time, Orange Island had grown northward and westward, and the Suwannee Strait had become much narrower and shallower. The coral reef complexes off the western coast of Orange Island had also expanded by this time, forming a lagoon system and chain of coral islands similar to the Recent Florida Keys. The carbonate banks of the Everglades region had also enlarged and had become shallower. Off the western side of Orange Island, upwelling systems were beginning to form, ushering in the Upwelling-Deltaic Depositional Episode.

Figure 1.5 — The **Arcadia Subsea** (Burdigalian to early Serravallian Miocene). For the first time since its inception, the Floridian Peninsula became a continuous land mass during the time of the Arcadia Subsea (see Chapter 2). The Suwannee Strait was essentially closed by this time, being filled with dense mangrove forests. In the south, the Everglades area still remained a deep lagoon surrounded by a series of banks. The largest and widest of these was along the southwestern coast and was made up of extensive oyster bars and ahermatypic coral bioherms. The upwelling systems along western Florida, established during Tampa time, had become permanent features by the Burdigalian Miocene. These produced immense phytoplankton blooms all along the coast which, in turn, poured into the Arcadia Subsea basin. The rich phosphate deposits of the lower Hawthorn Group are a result of the accumulation of these upwelling-driven plankton resources.

Figure 1.6 — The **Polk Subsea** (late Serravallian to early Tortonian Miocene). The mid-Miocene was a dynamic time for the Everglades and southern Florida (see Chapter 3). The southwestern edge of the peninsula (the present DeSoto Plain) had developed into a broad, shallow coastal lagoon system. The DeSoto Plain area itself was undergoing massive karstic wasting at this time, developing large sinkhole lakes and extensive river systems. The sediments carried down from the DeSoto Plain fed into the carbonate bank system along the western side of the Polk Subsea, producing a wide area of shoals. Extensive oyster bioherms formed along the seaward edge of the lagoon system and on the banks farther south. Upwelling systems along western Florida became stronger at this time, flooding the western coastal lagoon systems and the Okeechobean basin with nutrient-rich water and massive phytoplankton blooms. These upwelling-driven resources ultimately produced the phosphates of the upper Hawthorn Group. Along the eastern side of the Okeechobean Sea, the carbonate banks remained narrow and undeveloped.

Figure 1.7 — The **Charlotte Subsea** (late Tortonian to early Messinian Miocene). The river and lake systems that formed along the DeSoto Plain area reached their maximum development during this time. The erosion caused by this interconnected fluvial and lacustrine system (primarily in the southern area of the Mount Dora Ridge and Lake Wales Ridge) produced a huge sediment budget that poured southward into the Okeechobean Sea basin. The oyster banks along the western side of the Charlotte Subsea deflected the quartz sand and clay sediments southward and eastward, producing the DeSoto Delta (see Chapter 3). This giant clastic wedge quickly filled in the western half of the Okeechobean basin, producing a broad, shallow platform. The upwelling systems off western Florida continued throughout this time, as did cooler water temperatures. This temperate marine climate allowed many Chesapeake Miocene mollusks, previously known only from the Eastover Formation of Virginia and northern North Carolina, to invade the Charlotte Subsea.

Figure 1.8 — The **Murdock Subsea** (early Zanclean Pliocene). Following the late Messinian Miocene sea level drop and worldwide cooling event (the "Messinian Event"), the Floridian Peninsula again underwent a time of accelerated erosion (see Chapter 3). The sediments from this erosional time poured into the Everglades area, producing a second giant clastic wedge, the Immokalee Delta. By the beginning of the early Pliocene, the Okeechobean Sea reflooded the area, drowning the Immokalee Delta and distributing its quartz sand and clay sediments at least two thirds of the distance across the entire Everglades Basin. This left only a narrow, crescent-shaped deep water area along the eastern side of the Murdock Subsea. The upwelling system along western Florida persisted into this time, again producing cool, nutrient-rich water and continuous plankton blooms. This phytoplankton resource supported immense bioherms of giant barnacles and oysters, and a large fauna of marine mammals, including whales, seals, dugongs, and walruses. During the sea level drop at the end of the Zanclean Pliocene, the western Floridian upwelling systems ceased and phosphorite production came to an end. The Murdock Subsea represents the last stage of the Upwelling-Deltaic Depositional Episode and the end of Hawthorn Group deposition.

Figure 1.9 — The **Tamiami Subsea** (late Zanclean to late Piacenzian Pliocene). The warm tropical conditions of the mid-Pliocene, and the cessation of permanent upwelling systems, allowed the Okeechobean Sea to return to an episode of carbonate deposition. The platform produced by the DeSoto and Immokalee Deltas now became covered by a thick layer of carbonate sediment, producing an environment similar to the Recent Bahama Banks (see Chapter 4). Dominating the Tamiami Subsea was the Everglades Pseudoatoll, a "U"-shaped system of zonated coral reefs and coral islands that encircled the Everglades Basin. These amazingly rich coral assemblages (over 100 species) grew on top of the oyster and barnacle banks that were produced during Hawthorn time. Along with the coral complexes, extensive mangrove jungles and tropical estuaries also existed along the northern edge of the pseudoatoll (along the mainland) and extended into the Kissimmee Embayment. Never before in the history of the Everglades area had there been such a wide diversity of habitats and depositional environments, and this led to the evolution of the richest-known American molluscan and coral faunas (beautifully preserved in the Tamiami Formation). The incipiency of the Everglades Pseudoatoll marks the beginning of the final episode of Everglades sedimentary deposition.

Figure 1.10 — The **Caloosahatchee Subsea** (late Piacenzian Pliocene to late Calabrian Pleistocene). The second phase of the Pseudoatoll Depositional Episode took place during the transgressive sequences of the Caloosahatchee Subsea (see Chapter 5). The geomorphologic template put down by the Tamiami pseudoatoll was expanded upon during this time, with the intralagoonal features becoming wider and shallower. Mangrove forests covered large areas of the western side of the Okeechobean Sea and the entire southern coast of the Floridian Peninsula. At this time, the first major barrier island system (Nashua Lagoon System) began to form along the entire eastern coast of Florida.

Figure 1.11 — The **Loxahatchee Subsea** (Aftonian Pleistocene). The final depositional filling of the Okeechobean Sea began with the flooding of the Loxahatchee Subsea (see Chapter 6). The lagoonal area of the pseudoatoll was greatly reduced, with over half the area filled with mangrove forests and islands. The Reticulated Coastal Swamps area of Recent southwestern Florida is the last remnant of this vast mangrove island system. The narrow reef tracts along the eastern side of the Okeechobean Sea had enlarged to become wide islands that were covered with pine and palmetto forests. The Kissimmee Valley and Okeechobee Plain areas were also filled with dense mangrove forests. Only the eastern side of the Okeechobean Sea, adjacent to the eastern islands, contained deep water areas. This landlocked lagoon remained warmer than the surrounding ocean and acted as a refugium for tropical mollusks and corals that had become extinct elsewhere along the eastern U.S. coastline. During the subsequent Kansan Glacial Stage, the Loxahatchee Subsea retreated and the deep lagoon area became a large freshwater lake (Lake Okeelanta).

Figure 1.12 — The **Belle Glade Subsea** (Yarmouthian Pleistocene). Enlarging upon the geomorphologic features and bathymetric contours of the Loxahatchee Subsea, the Belle Glade Subsea was the last oceanographic subdivision to retain the basic structure of the pseudoatoll. By this time, mangrove forests and islands filled at least three quarters of the basin and lined the adjacent Florida mainland. The western banks and reefs had given rise to a large island covered with pine and palmetto forests (Immokalee Island). Similarly, the eastern island chain, particularly the southern part, was covered by dense pine and palmetto forests. The main lagoon of the Okeechobean Sea was now much shallower, with only a small area in the center containing any deep water areas (see Chapter 6).

Figure 1.13 — The **Lake Worth Subsea** (late Illinoian to late Sangamonian Pleistocene). This last major marine flooding of the Everglades set the stage for the geomorphology of Recent southern Florida (see Chapter 7). The eastern lagoon of the Okeechobean Sea was now filled with carbonate and quartz sand sediments. A long chain of coral keys had now formed along the southeastern edge of the continental shelf and extended westward to present-day Bahia Honda Key in the Florida Keys. The Bahamas Banks were submerged at this time and large oceanic swells passed unabated from West Africa directly to the Florida coast. These heavy surf conditions along the southeastern reef systems produced large amounts of oölite which, in turn, washed behind the reef tracts and accumulated in wide banks. The Lake Worth Subsea oölite banks extended from near present-day Fort Lauderdale and Miami, southward across Dade and Monroe Counties and Florida Bay to offshore of Key West.

Figure 1.14 — **Recent Florida** (Holocene). This now-famous NASA space shuttle photograph of the Floridian Peninsula clearly shows the Everglades and the higher features (with cities) on either side. Although subtle, the outline of the predecessor pseudoatoll is still apparent. The Everglades truly is the final landscape of the Okeechobean Sea. (Image courtesy of the Image Science and Analysis Laboratory, NASA Johnson Space Center; photograph reference STS 51c-44-25.)

2 Late Paleogene to Early Miocene Southern Florida

Looking at Florida today, it is hard to visualize how utterly different the state appeared just 40 million years ago during the late Eocene Epoch. At that time (Jackson Stage), what would become Florida was a small, shallow carbonate bank over 150 km off the coast of Georgia (Figure 2.1). Resembling the Recent Bahamas, this shallow area, the **Ocala Bank** (named for the Ocala Group of formations), was separated from the U.S. mainland by a wide, shallow seaway, the **Suwannee Strait** (named for the Suwannee Limestone). On the Ocala Bank and on the shallow areas of the surrounding Florida Platform, biogenic carbonate sediments rapidly accumulated, forming thick layers. As the bank grew over the time span of the Eocene Epoch, six formations were deposited: the oldest, the Oldsmar Formation (Wilcox Stage); the middle (Claiborne Stage) Lake City and Avon Park Formations; and the youngest (Jackson Stage), the Ocala Group, with its Inglis, Williston, and Crystal River Formations.

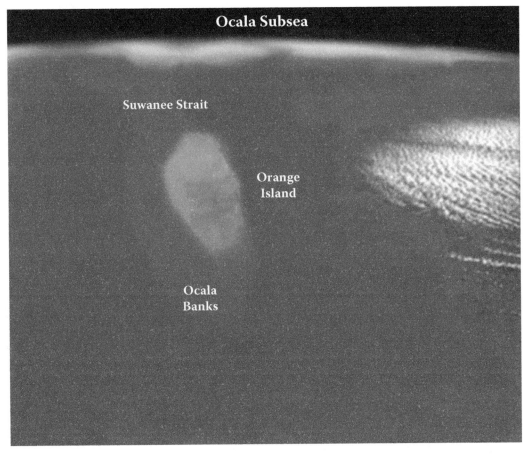

FIGURE 2.1 Simulated space shuttle image (altitude: 100 mi) of the Floridian Peninsula during the Priabonian Eocene, showing the possible appearance of the Ocala Sea and some of its principal geomorphologic features.

During Jackson time, the tropical environments in the area around the Ocala Bank and the Suwannee Strait had produced enormous quantities of carbonate sediments, creating neritic oceanographic conditions that were unique in the eastern U.S. This area, here referred to as the **Ocala Sea**, differed greatly from the contemporaneous adjacent areas to the north, in the Mississippi Embayment, and along the Atlantic Coastal Plain. Instead of being influenced by continental siliciclastic sediments, the Ocala Sea retained pristine, high-tropical carbonate environments.

THE PALEOGEOGRAPHY OF THE OCALA SEA

The principal paleogeographic feature of the Ocala Sea was the Ocala Bank, and this formed the core of the incipient Floridian Peninsula. Geomorphologically complex, the Ocala Bank was covered with small coral cays, probably in a linear form like the island chains of the Recent Great Bahama Bank. These island chains later fused together to form a single large island (Orange Island, described later in this chapter). The protected carbonate flats within the central lagoon were of special interest in that they housed some of the first Turtle Grass (*Thalassodendron*) beds. This important sea grass had just evolved in the early Eocene and made its first appearance in Florida in the Avon Park Formation. The first mangrove trees were also evolving at this time, and the coral cays of the Ocala Bank probably supported forests of these pioneer species. The coral reefs, containing many genera of corals that are still extant, along with the Turtle Grass and mangrove forests, would have given the Ocala Bank a distinctly modern ecological appearance.

As mentioned earlier in Chapter 1, the northern end of the Ocala Bank would have been in the vicinity of present-day Gainesville, Alachua County, and the southern end would have been in the area of Orlando, Orange County. South of the Orlando area, the bank sloped gradually, at about a 10° angle, to the area of the present-day Florida Keys. The slope south of the Ocala Bank was covered by thick deposits of carbonate fines that had eroded off the coral reefs and carbonate lagoon areas to the north. These lime muds gave rise to the three formations of the Ocala Group, the last deposition of the Eocene. In late Jackson time, the sediments of the three formations of the Ocala Group would have been unconsolidated, or only partly consolidated, and would have been draped over the older indurated limestones of the Avon Park Formation. The region that would eventually be the Everglades would have encompassed a small area on the extreme southeast corner of the Florida Platform and would have been at a depth of around 300 m. A similar environment is seen in the deeper areas of the present-day Straits of Florida, where carbonate sand and mud cascade down over older limestone terraces.

MEGATSUNAMIS AND THE EVERGLADES UNCONFORMITY

The end of the Eocene has long been known to have been a time of catastrophic environmental conditions and an accompanying mass extinction, now thought to have been caused by an asteroid impact (Ganapathy, 1982; Maurasse and Glass, 1976; Sanfilipo et al., 1985). Based on the chemical composition and distribution of Eocene microtektites in the Caribbean Sea and Gulf of Mexico (Glass and Zwart, 1979), it had been determined that the asteroid impact occurred somewhere along the eastern U.S. Finally, in the late 1990s, the impact site was discovered at the mouth of Chesapeake Bay (Poag, 1999; Powars and Bruce, 1999; Powars, 2000; Poag et al., 2004). When the force of the impact and the size of the Chesapeake Bay Crater were finally reconstructed, it was found that the resultant series of catastrophic events affected the entire Eastern seaboard, including the Ocala Bank and the Florida Platform.

The asteroid that produced the Chesapeake Bay Crater, as impressive as it was, judging from the size of its astrobleme, was actually part of an even larger object. Entering the Earth's atmosphere over the North Pole, this asteroid broke into two large fragments and innumerable smaller pieces.

The largest fragment hit the continental shelf off Virginia (the present-day mouth of Chesapeake Bay), whereas the next largest fragment hit Popigai, Siberia (Bottomley et al., 1997). The debris field from the American fragment, alone, extended from Chesapeake Bay to off southern Argentina, with a width of over 300 km. The entire eastern U.S. was peppered with millions of fragments, both from the shattered asteroid and from the impact ejecta debris cloud. The main crater, itself, was over 80 km in diameter at impact. The disintegrating asteroid fragment produced not only an initial blast-splash of over 40 km in height, but also tremendous steam explosions and a wide zone of concentric fractures. Today, the resultant impact structure is over 128 km in diameter (Poag, 1999). Numerous suspected astroblemes in the Carolinas, such as Lake Waccamaw, may prove to have been formed by smaller fragments from the disintegrating larger body.

Although the actual impact event was catastrophic in the extreme, what followed had an even greater effect on southern Florida. Within minutes of striking the shallow sea floor off Virginia, the disintegrating asteroid would have generated a series of megatsunamis, with some reaching over 300 m in height (Poag, 1999). These would have traveled down the entire eastern coast of the U.S. and, hours later, would have passed over the Ocala Bank and into the Gulf of Mexico. The force of these giant waves would have scoured the entire Florida Platform, removing most of the unconsolidated seafloor sediments. The shallow slope areas east and southeast of the Ocala Bank would have been the first places to be exposed to the full impact of the megatsunamis and would have been particularly vulnerable to scouring. This catastrophic event is preserved under the Everglades in the form of a large unconformity (Figure 2.2). Although known to geologists for over two decades (Merritt et al., 1983: 29, Figure 5; Missimer, 1984: 388), these missing sediments, here referred to as the **Everglades Unconformity**, were originally thought to have resulted from some form of postdepositional erosion. In the area of the unconformity, the entire Ocala Group is missing, with the continuously distributed younger (Oligocene) Suwannee Limestone lying directly on top of the Claiborne-aged Avon Park Formation. Considering that the asteroid impact took place at the end of the Ocala deposition, the Everglades Unconformity is chronologically perfectly situated and may be the first record of impact-related catastrophism south of the Chesapeake Bay area.

Geophysical studies of the Everglades Unconformity area have recently uncovered two interesting anomalies. These are shown in Figure 2.3, superimposed upon the surface contours of the Eocene formations. The first, and most intriguing, is a small magnetic anomaly located within a valley-like depressional feature under the southern Everglades. This strong magnetic anomaly, surrounded by a steep magnetic gradient, was first shown on a magnetic survey map of the Floridian Peninsula produced by Klitgord et al. (1984) in their study of Jurassic transform faults. This isolated, geographically small anomaly, with a field of 100 to 250 nT, stands out strikingly from the surrounding low-magnetic area of only 0 to 10 nT. As can be expected on a thick carbonate platform devoid of igneous rocks, the low-magnetic area extends for over 300 km to the south and west of the magnetic high and over 200 km to the north and east. The presence of a well-documented, lone, high-magnetic anomaly, coincident with a valley-like depression in the Eocene rocks, is indeed a most interesting feature.

The second anomalous geophysical feature associated with the Everglades Unconformity is a pair of large gravity lows, which were also illustrated by Klitgord et al. (1984) on a gravity survey map from their Jurassic study (shown here on Figure 2.3). These two lows, with fields of less than –5 mG, appear to be related to "high transmissivity zones" or "cavity zones" that are found in the early and middle Eocene, Paleocene, and Cretaceous rocks of the southern tip of the state. These labyrinthine vug structures (referred to as "boulder zones" by drillers) and interconnected cavities may have formed by subsequent dissolution enlargement of brecciated limestone. With the exception of the Everglades area, the older Eocene formations (Oldsmar and Avon Park), across the State, characteristically contain gypsiferous limestones and anhydrite beds (diagenetically produced). These anhydrites are especially well developed under the central and northern parts of the peninsula.

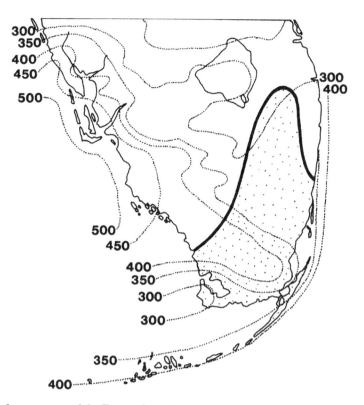

FIGURE 2.2 Surface contours of the Eocene formations below the Everglades region. The stippled section delineates the area where the late Eocene (Jackson Stage) Ocala Group is missing, and the middle Eocene (Claiborne Stage) Avon Park Formation is exposed. The contours are in meters below mean surface. The area of the missing Ocala Group is referred to here as the Everglades Unconformity. (Taken from Merritt, M.I., Myer, F.W., Sonntag, W.H., and Fitzpatrick, D.J., 1983, *Subsurface Storage of Freshwater in South Florida: A Prospectus*, U.S. Geological Survey Water Resources Investigations Report 83–4214, pp. 29–40.)

Under the Everglades, there is a distinct shift in lithofacies, with the anhydrite beds being replaced by fractured and vuggy dolostone–limestone facies (Merritt et al., 1983; Puri and Winston, 1974). These "boulder zone" fracture systems follow curving, almost concentric, patterns and house migratory petroleum reserves. The main oil-producing fields on peninsular Florida, the Corkscrew and Sunniland fields, correspond to these curving deep fractures and fault systems along the western and southwestern sides of the Everglades (Applegate, 1986; McCaslin, 1986). Recent investigations on deep-injection well possibilities by private engineering firms (unpublished; personal communication) have shown that these curving fracture systems may also extend around the eastern side of the Everglades (corresponding to the gravity low shown on Figure 2.3).

The coincidence of the Everglades Unconformity and the underlying geophysical and geologic anomalies is compelling. These data lead to several possible scenarios, all of which, unfortunately, must remain in the realm of theory for the present time. Considering the new discovery of the Chesapeake Bay asteroid impact, it seems that the two most likely scenarios involve either the effects of the megatsunamis themselves or the megatsunamis coupled with an impact by a small debris fragment. The first scenario explains the missing seafloor of the Everglades Unconformity and also the gravity lows–brecciated limestone patterns (as shattered debris reworked in the tsunami surges).

The second scenario, first postulated by the senior author (Petuch, 1987), has the extreme southeastern edge of the Florida Platform being struck by a small impactor, producing a crater-like feature on the seafloor (in 300 m water depth), and shattering the underlying limestones to a depth

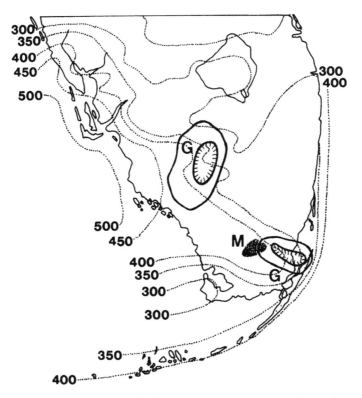

FIGURE 2.3 Geophysical anomalies below the Everglades area, superimposed upon the surface contours of the Eocene formations. Two gravity lows (G) of −5 mG (milliGals) can be seen to be associated with a magnetic high (M) of 100–250 nT (nanoTeslas). The magnetic anomaly lies in a valley-like feature, and the gravity lows are associated with fracture systems and areas of cavity zones. (Taken from Klitgord, K.D., Popenoe, P., and Schouten, H., 1984, Florida: A Jurassic transform plate boundary, *Journal of Geophysical Research,* 89: 7753–7772.)

of 8 km. When the newly discovered megatsunamis are also factored into this model, the formation of all the anomalous Everglades features can be combined into a single catastrophic event. Future research within the Everglades may reveal the true nature of the magnetic high, the adjacent gravity lows, and the valley-like structure. Considering the size and density of the Chesapeake impact debris field, geologists should be on the lookout for other small impact structures in the Eocene rocks on the Floridian continental shelf in the Gulf of Mexico and in Georgia and northern Florida.

PALEOGEOGRAPHY OF THE DADE SUBSEA, OKEECHOBEAN SEA

Immediately following the late Eocene asteroid impact, the world was plunged into a severe cold time, with expansive ice buildup in the Northern Hemisphere. This led to a major sea level drop, with eustatic lows reaching 200 to 300 m below present sea level. From the latest Eocene (36 million years) until the mid-Rupelian Oligocene (32 million years), southern Florida remained above sea level, with dry subtropical to warm-temperate conditions prevailing. During this 4 million-year terrestrial episode, the Avon Park Formation (in the Everglades) and the Crystal River Formation (elsewhere in Florida) were exposed as the land surface across the entire Florida Platform. In the area of the Everglades Unconformity, groundwater penetrated the underlying fractures, dissolving the smaller limestone fragments and enlarging the spaces between brecciated blocks. Over time, this dissolution caused the entire southern Everglades area to slowly collapse, forming a shallow

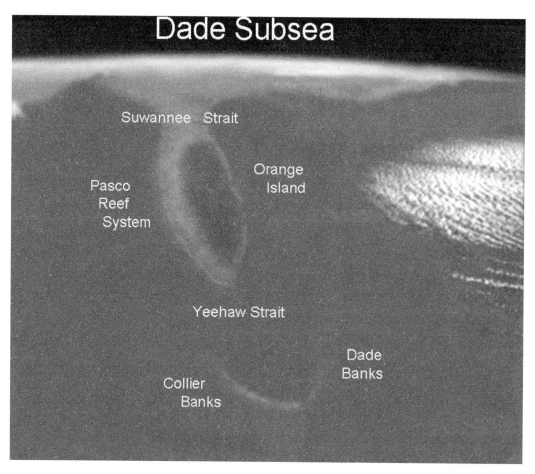

FIGURE 2.4 Simulated space shuttle image (altitude: 100 mi) of the Floridian Peninsula during the Rupelian Oligocene, showing the possible appearance of the Dade Subsea of the Okeechobean Sea and some of its principal geomorphologic features.

depressional feature. This dissolutional collapse feature may also have formed as an enlarged version of the older late Eocene valley structure. Whatever its means of formation, the southern Everglades depression is an obvious feature and is readily demonstrated in the stratigraphic columns of Puri and Winston (1974). In their data, the top of the younger Oligocene Suwannee Limestone is seen to conform to a 100 m-deep depression.

By the beginning of mid-Rupelian time, the worldwide climate began to warm and sea levels rose almost to their late Eocene levels. A shallow sea, the Dade Subsea (Figure 2.4), now flooded the southern Florida Platform and produced depositional environments that resulted in the sediments of the Suwannee Limestone (Everglades component). The slightly lower sea levels of the early Oligocene allowed the remnants of the Ocala Bank to remain above water, forming a large, low-relief island (**Orange Island**, named for Orange County; see Vaughan, 1919). Because of lower sea levels, the Suwannee Strait was now much shallower than during the Eocene and contained extensive coral reef systems. These reefs extended northward, into southern Alabama (the Salt Mountain Formation) and southern Georgia (the Flint River Formation), and southward, forming a large reef tract along the western side of Orange Island (the **Pasco Reef System**, named for Pasco County; referred to as the "Orange Island Reef System" by Petuch (2004: 57). These western Orange Island reefs housed amazingly rich and endemic coral and molluscan faunas. Many of these are now found living only in the South Pacific. The entire area more closely resembled Recent northern Australia than it did the present-day Caribbean (see Petuch, 1997). The southern Alabama, southern Georgia, Suwannee Strait, and northern Orange Island area was included within the Bainbridge Subsea of the Choctaw Sea (see Petuch, 2004: 2–5). Only the area south of Orange Island constituted the Dade Subsea of the Okeechobean Sea.

At the extreme southeastern end of the Florida Platform, two arc-shaped carbonate banks had formed around the higher edges of the southern Everglades depression. These two submerged structures, the **Dade Banks** (named for Dade County) and the **Collier Banks** (named for Collier County), created the template for all subsequent reefal deposition in the Everglades area. The Suwannee Limestone sediments from these bank areas are composed of unconsolidated calcarenites and thick beds of foraminiferal tests (Missimer, 1984: Figure 6). Separating the depressional feature and its arc-shaped banks from Orange Island was a wide, deep channel, the **Yeehaw Strait** (named for Yeehaw Junction, Osceola County, which lies at its center). By the end of Suwannee time (early Chattian Oligocene), the Yeehaw Strait and Everglades depression began to fill with carbonate fines that were eroding off Orange Island and the Pasco Reef System. The first faint outline of the modern Floridian Peninsula was beginning to emerge. The paleoenvironments contained in the Suwannee Limestone were described by the senior author (Petuch, 2004: 57–67).

THE SUWANNEE LIMESTONE, VICKSBURG GROUP

HISTORY OF DISCOVERY

Although known for decades from exposures along the Suwannee River and from quarries in north central Florida, the Suwannee Limestone was not formally described until 1936, when Cooke and Mansfield proposed the name Suwannee. Because of its commercial importance as a local source of railway ballast and as an aggregate for concrete (Cooke, 1945: 90–91), the formation had been chemically analyzed over a decade before it was formally named (Mossom, 1925). The Suwannee is still quarried as a source of limestone aggregate in Citrus, Hernando, Pasco, and Polk Counties.

LITHOLOGIC DESCRIPTION AND AREAL EXTENT

As described by Cooke (1945: 87; condensed here):

> Where the Suwannee is unaltered. It consists of a soft granular mass of limy particles, many of them of organic origin. The color is commonly yellow or cream, locally with a pinkish tinge. At many surface exposures all the lime has been leached from it, leaving a porous or massive flint, which is recognizable as Suwannee by the presence of molds of the common Suwannee echinoid, *Cassidulus gouldii* [authors' note: now *Rhyncholampas gouldii*] … . Chemical analyses show that the Suwannee limestone contains about 91–98% calcium carbonate and that the chief impurity is silica.

The lithology of the Suwannee Limestone in the Everglades was best summarized by Missimer (1984: 388–389) (condensed here):

> Lithologically, the Suwannee is predominantly a light-colored limestone, which commonly contains soft calcarenite consisting of foraminifera tests. It does contain some unlithified lime muds and well-indurated dolomites. The dolomites are commonly hard, thin, and interbedded with limestones. Some unlithified dolosilts occur in the middle part of the stratigraphic section … .

The depositional area of the Suwannee conforms to the paleogeography of the Dade Subsea. In northern and western peninsular Florida, the formation follows the contours of the coral reef systems and carbonate lagoons of the Suwannee Strait and Pasco Reef System. These limestones often contain beautifully preserved siliceous pseudomorphs (agatized) of an especially rich mollusk and coral fauna (examples are shown here in Figure 2.5 and Figure 2.6). Specimens of these

Suwannee agatized fossils have been found as far south as DeSoto County. South of there, these macrofossils are replaced by microfossils, mostly foraminifera, reflecting the deep water conditions of the Oligocene Everglades. Stratigraphically, the Suwannee appears to be more complex than originally thought, containing at least three distinct units. Future research may show that these constitute members yet to be described.

The Suwannee Limestone underlies the entire area that is covered by this book. As mentioned earlier, the formation has a conformable contact with the underlying late Eocene Ocala Group throughout the region, with the exception of the Everglades area. There, the Suwannee lies unconformably on the middle Eocene Avon Park Formation. The thickness of the Suwannee in southern Florida is variable, with a maximum of 180 m under Lee County and a minimum of 18 m under Dade County. The general surface contours of the Suwannee show a gradual dip to the south, being approximately 100 m below the surface in DeSoto County, 250 m in Palm Beach County, 210 m in Lee County, 300 m in southern Palm Beach County and Broward County, and over 400 m in the depressional feature in western Dade and northernmost Monroe Counties. South of this feature, in Monroe County (including Florida Bay and the Florida Keys), the Suwannee Limestone comes closer to the surface, at around 200 m below the northern Keys (Puri and Winston, 1974).

STRATOTYPE

From Cooke (1945: 86): " … The name 'Suwannee Limestone' was proposed by Cooke and Mansfield (1936) for a yellowish limestone typically exposed along the Suwannee River, from Ellaville almost to White Springs." A typical section was illustrated by Cooke (1945: 88, Figure 12; a photograph by Herman Gunter) from the left bank of the Suwannee River " … about half a mile above the old bridge on the road from White Springs to Lake City" (now U.S. Highway 41), Hamilton County.

AGE AND CORRELATIVE UNITS

Based on the presence of many classic Mississippi Vicksburg Group fossil mollusks, the Suwannee Limestone is now known to span the period from the early middle Rupelian Oligocene to the earliest Chattian Oligocene, a range of over 5 million years. Some of the more important of these classic Mississippi Oligocene fossils found in the Suwannee include *Clavolithes vicksburgensis*, *Conorbis porcellanus*, *Ficus mississippiensis*, *Torcula mississippiensis*, *Talityphis mississippiensis*, and *Terebrellina divisura*. These typical Vicksburg index fossils and many others were reported from the Suwannee and illustrated for the first time by the senior author (Petuch, 1997; 2004: 57–67). The Suwannee Limestone of the Everglades region is exactly correlative with the Byram Formation of the Vicksburg Group of Mississippi, the Flint River Formation of southern Georgia, and the Salt Mountain Formation of Alabama.

SUWANNEE INDEX FOSSILS

Along with the Mississippi Vicksburg Group species, the Suwannee molluscan fauna contains a large endemic component that can be used to demarcate the formational boundaries. These endemic Floridian taxa are representative of the Hernandoan Mollusk Age. Some of the more easily recognizable species include:

Gastropoda
Astraea polkensis
Apicula bowenae (Figure 2.6B)
Pyrazisinus kendrewi
Telescopium blackwaterense (Figure 2.5L)

Telescopium hernandoense
Prismacerithium prisma (Figure 2.6J)
Cestumcerithium brooksvillensis (Figure 2.5J)
Orthaulax hernandoensis (Figure 2.5D)
Amauropsis mansfieldi (Figure 2.6K)
Terebellum hernandoensis
Pterynotus propeposti (Figure 2.5K)
Spinifulgur gemmulatum
Turbinella suwannensis (Figure 2.5F)
Vasum suwanneensis (Figure 2.5G)
Omogymna brooksvillensis (Figure 2.5B)
Persicula suwanneensis (Figure 2.6C)
Falsilyria kendrewi (Figure 2.6L)
Hermes kendrewi (Figure 2.5E)
Suwannescapha lindae (Figure 2.6D)
Bivalvia
Glycymeris suwannensis (Figure 2.6H)
Dimarzipecten brooksvillensis
Trachycardium brooksvillensis (Figure 2.6A)
Lucina perovata
Hippopus gunteri (the only known American member of the Tridacnidae)

The Suwannee fauna also shares many species with the Flint River Formation of Georgia. These all appear to have been associated with the Rupelian Oligocene coral reef systems of the Suwannee Strait and Pasco Reef System areas. The senior author has previously described and illustrated most of these shared taxa (Petuch, 1997; 2004), and some of the more easily recognizable include:

Gastropoda
Cerithioclava eutextile
Cerithium cookei
Ampullinopsis flintensis
Hyalina silicifluvia
Falsilyria mansfieldi (Figure 2.5H)
Gradiconus cookei
Protoconus vaughani
Bivalvia
Chione bainbridgensis (Figure 2.6I)

Of particular importance as an index fossil for the Suwannee Limestone, as pointed out by Cooke (1945), is the echinoid *Rhyncholampas gouldii* (Figure 2.5C) (originally placed in the genus *Cassidulus*). This small echinoid is the most conspicuous and widespread fossil in the Suwannee Limestone. In the reefal facies of the Suwannee, a large fauna of corals can also be used to delineate the formational boundaries. Many of these corals are shared with the Flint River Formation of Georgia and the Salt Mountain Formation of Alabama. Others are shared with the Antigua Formation of the Lesser Antilles (Caribbean Oligocene). Some of the more commonly encountered corals include:

Cnidaria–Scleractinia
Acropora panamensis
Acropora saludensis
Stylophora minutissima (Figure 2.6F)

FIGURE 2.5 Index fossils for the Suwannee Limestone, Vicksburg Group. A = *Cypraeorbis kendrewi* Petuch, 1997, length 26 mm; B = *Omogymna brooksvillensis* (Mansfield, 1937), length 20 mm; C = *Rhyncholampas gouldii* (Bouvé, 1846), length 42 mm; D = *Orthaulax hernandoensis* Mansfield, 1937, length 62 mm; E = *Hermes kendrewi* (Petuch, 1997), length 33 mm; F = *Turbinella suwannensis* (Mansfield, 1937), length 107 mm; G = *Vasum suwanneensis* Petuch, 1997, length 45 mm; H = *Falsilyria mansfieldi* (Dall, 1916), length 32 mm; I = *Melongena (Myristica) crassicornuta* Conrad, 1848, length 49 mm; J = *Cestumcerithium brooks-villensis* (Mansfield, 1937), length 24 mm; K = *Pterynotus propeposti* (Mansfield, 1937), length 35 mm; L = *Telescopium blackwaterense* (Mansfield, 1937), length 43 mm. All specimens are siliceous pseudomorphs, replaced by agate.

FIGURE 2.6 Index fossils of the Suwannee Limestone, Vicksburg Group. A = *Trachycardium brooksvillense* (Mansfield, 1937), length 28 mm; B = *Apicula bowenae* (Mansfield, 1937), length 39 mm; C = *Persicula suwanneensis* Petuch, 1997, length 10 mm; D = *Suwannescapha lindae* Petuch, 1997, length 18 mm; E = *Astrocoenia decaturensis* Vaughan, 1919, length 23 mm; F = *Stylophora minutissima* Vaughan, 1900, length 23 mm; G = *Solenosteira suwanneensis* Petuch, 1997, length 27 mm; H = *Glycymeris suwannensis* Mansfield, 1937, length 15 mm; I = *Chione bainbridgensis* Dall, 1916, length 26 mm; J = *Prismacerithium prisma* Petuch, 1997, length 19 mm; K = *Amauropsis mansfieldi* (Petuch, 1997), length 28 mm; L = *Falsilyria kendrewi* Petuch, 1997, length 36 mm. All specimens are siliceous pseudomorphs, replaced by agate.

Stylophora ponderosa
Goniopora decaturensis
Actinactis alabamensis
Astrocoenia decaturensis (Figure 2.6E)
Montastrea bainbridgensis
Antiguastrea silicensis

Many of these taxa were illustrated previously, and their biogeographical distributions discussed, by the senior author (Petuch, 2004: 25, 62, 65 to 67). Some of these species may also be present in the area of the Collier and Dade Banks, and drillers working in these areas should be aware of their possible occurrence. The marine communities of the Suwannee Limestone were also described by the senior author (Petuch, 2004: 57 to 67) and include the *Telescopium hernandoense* Community (mud flat and mangrove forest environments), the *Lucina perovata* Community (Turtle Grass bed environment), the *Orthaulax hernandoensis* Community (shallow carbonate sand bottom environment), the *Stylophora minutissima* Community (back reef environment), and the *Goniopora decaturensis* Community (reef platform environment).

PALEOGEOGRAPHY OF THE TAMPA SUBSEA, OKEECHOBEAN SEA

During the middle Oligocene (early Chattian Age), the world climate again cooled, and this resulted in a sea level drop that exposed the entire Dade Subsea, Orange Island, and the Suwannee Strait area to dry land conditions. After over 1 million years of terrestrial environments, the Okeechobean Sea again reflooded the Everglades area, this time as the Tampa Subsea (named for Tampa, Hillsborough County) (Figure 2.7). During the late Chattian Oligocene and Aquitanian Miocene, Orange Island had enlarged both northward and westward (to present-day DeSoto County), with a large tropical carbonate lagoon system having formed along the western side (the **Hillsborough Lagoon System**, named for Hillsborough County). Fringing the western edge of the lagoon system were the two principal geomorphologic features of the Tampa Subsea: the **Tampa Reef Tract**, a zonated coral reef system, and the **Tampa Archipelago**, a linear chain of small coral cays resembling the Recent Bahamian Islands. The first extensive permanent upwelling systems began at the end of Aquitanian time, with phosphate-rich plankton blooms being washed into the Hillsborough Lagoon System and Okeechobean Sea basin.

To the south of Orange Island and the Tampa Archipelago, across the Yeehaw Strait, the Everglades Basin still remained a deepwater area, with depths of over 250 m along the eastern side. The Collier and Dade Banks also had enlarged by the Aquitanian Miocene, becoming wider and extending farther northward. In the north, the Suwannee Strait and adjacent areas in Georgia (the Chattahoochee Subsea of the Choctaw Sea; Petuch, 2004) had begun to fill in, becoming shallower and narrower. Extensive mangrove forests were established over much of the Chattahoochee Subsea at this time (the **Coosawhatchee Lagoon System**), contributing to the increased shoaling. The Tampa Reef Tract, Hillsborough Lagoon System, and the Everglades Basin served as the centers of deposition for the Nocatee and Tampa Members of the Arcadia Formation (Scott, 1988). These basal members represent the beginning of the Hawthorn Group.

THE ARCADIA FORMATION, HAWTHORN GROUP

First formally proposed by Scott (1988), the Arcadia Formation exhibits complex depositional patterns, both geographically and chronologically, and spans the time frame of two different Okeechobean subseas (the Tampa Subsea and the Arcadia Subsea). Only the mid-Pleistocene Bermont Formation exhibits this same type of depositional chronology, having members in two different subseas (the Loxahatchee Subsea and the Belle Glade Subsea). All the other formations

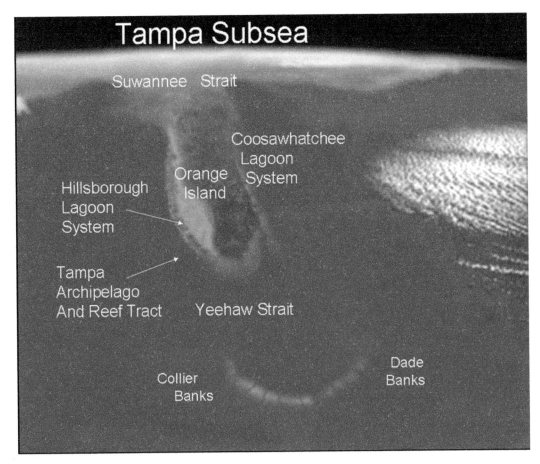

FIGURE 2.7 Simulated space shuttle image (altitude: 100 mi) over the Floridian Peninsula during the early Aquitanian Miocene, showing the possible appearance of the Tampa Subsea of the Okeechobean Sea and some of its principal geomorphologic features.

in the Everglades region are restricted to a single subsea. The Arcadia Formation contains the basal Nocatee, Tampa, and lower Arcadia Members and an upper Type Arcadia.

In this section of the book, we have decided to give an overall description of the Arcadia Formation first and then discuss the stratigraphic subdivisions in context with their depositional histories.

LITHOLOGIC DESCRIPTION AND AREAL EXTENT

As described by Scott (1988: 56–58; condensed here):

> ... The Arcadia Formation, with the exception of the Nocatee Member, consists predominantly of limestone and dolostone containing varying amounts of quartz sand, clay, and phosphate grains. Thin beds of quartz sand and clay often are present scattered throughout the section. These thin sands and clays are generally very calcareous or dolomitic and phosphatic ... Dolomite is generally the most abundant carbonate component of the Arcadia Formation, except in the Tampa Member. Limestone is common and occasionally is the dominant carbonate type. The dolostones are quartz sandy to very porous (moldic porosity) and micro-to-fine crystalline. The dolostones range in color from yellowish-gray to light olive-gray. The phosphate grain content is highly variable, ranging up to 25% but is more commonly in the 10% range ... the limestones are typically a wackestone to mudstone with few beds of packstone. They range in color from white to yellowish-gray ... fossils are generally present only as molds in the carbonate rocks ... Clay beds occur sporadically in the Arcadia Formation ... They are

thin, generally less than 5-feet thick, and of limited areal extent … color of the clay ranges from yellowish-gray to light olive-gray … smectite, illite, palygorskite, and sepiolite comprise the clay mineral suite … chert is also sporadically present in the Arcadia Formation in the updip areas … in many instances, the chert appears to be silicified clays and dolosilts.

The Arcadia Formation underlies the entire area covered by this book and follows the general contours of the underlying Suwannee Limestone. As pointed out by Scott (1988: 60), the top of the Arcadia ranges from 146 m below surface in Monroe County to 30 m below surface in Polk County. In the area of the Everglades Unconformity and depression feature (Palm Beach and Martin Counties), the top of the Arcadia is greater than 250 m below surface.

STRATOTYPE

From Scott (1988: 56):

> … The Arcadia Formation is named for the town of Arcadia in DeSoto County, Florida. The type section is located in core W-12050, Hogan #1, DeSoto County (SE ¼, NW ¼, Sec. 16, T. 38 S., R. 26 E.), surface elevation 19 m, drilled in 1973 by the Florida Geological Survey. The type Arcadia Formation occurs between −30 m MSL to −159 m MSL.

AGE AND CORRELATIVE UNITS

The Arcadia Formation, as defined by Scott, ranges from the late Chattian Oligocene to the Langhian Miocene, with a 4 million-year hiatus between the upper and lower members. In Florida, the Tampa Member and Nocatee Member, and the lower Arcadia beds, are exact equivalents of the Penney Farms Formation and the Chattahoochee and St. Marks Formations of the Panhandle. The upper Type Arcadia is the exact equivalent of the Marks Head Formation of northern Florida, the Torreya Formation of the eastern Panhandle, and the Chipola (Burdigalian) and Oak Grove (Langhian) Formations of the Alum Bluff Group.

THE NOCATEE MEMBER, ARCADIA FORMATION

This member from the lowest part of the Arcadia Formation was deposited along the southeastern corner of Orange Island and within the eastern section of the Hillsborough Lagoon System. The high quartz sand and clay content of the Nocatee was probably the result of a series of rivers discharging into the headwaters of the lagoon. These rivers would have transported sand and clay from the higher elevations on Orange Island downslope and into the area of the Yeehaw Strait and Hillsborough Lagoon. These same river-borne sediments, mixed with carbonate erosional fines from the Tampa Reef System, would have poured into the deep Okeechobean basin, laying down the lower beds of the undifferentiated Arcadia. The sediments of the Nocatee Member demonstrate that large Orange Island river systems were draining into the Tampa Subsea during the Chattian Oligocene and Aquitanian Miocene. These same rivers, later in the Miocene, may have been the source of the great delta systems (see Chapter 3).

Lithologic Description and Areal Extent

From Scott (1988: 73; condensed here):

> … The Nocatee is a predominantly siliciclastic unit in the type core. This is a noticeable change from the remainder of the Arcadia Formation, including the Tampa Member, which are predominantly carbonates with variable percentages of included siliciclastics. The quartz sands in the Nocatee are typically coarse grained … and range in color from white to light olive-gray. Phosphate grain content is quite variable. In the type core, phosphate grain content is generally low (1 to 3%) with scattered beds

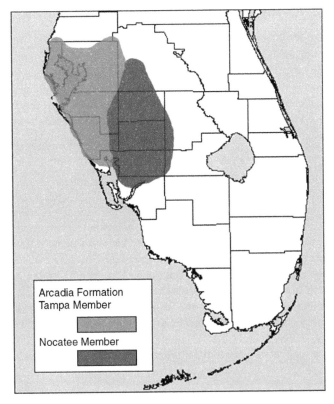

FIGURE 2.8 Areal distribution of the Nocatee and Tampa Members of the Arcadia Formation, Hawthorn Group.

with greater concentrations (up to 10%) … in the Nocatee Member in other cores, phosphate grains are more common, averaging 7 to 8% … Clay beds are quite common and are variably quartz sandy, silty, phosphatic, and calcareous to dolomitic. The colors characteristically range from yellowish-gray to light olive-gray and olive-gray. Limited x-ray data suggest that the characteristic clay mineral present is smectite, with palygorskite common. Illite and sepiolite are also present … Limestone and dolostones are both present in this member … The limestones are generally fine grained, soft to hard, quartz sandy and phosphatic … Colors of the limestones vary from white to yellowish-gray to olive-gray … The limestones are usually wackestones with varying degrees of recrystallization and cementation … The dolostones are quartz sandy, phosphatic, soft to hard. And micro-to-very finely crystalline … colors range from yellowish-gray to light gray, light olive-gray, and grayish-brown … Fossils are often present in the Nocatee, most often as molds."

The Nocatee Member is of limited areal extent (Figure 2.8), having been identified only in Polk, Hardee, DeSoto, Charlotte, Lee, Manatee, Hillsborough, Sarasota, and westernmost Glades and Highlands Counties. The top of the Nocatee ranges in depth from –24.5 m MSL in Polk County to –195 m MSL in Charlotte County.

Stratotype

From Scott (1988: 73):

… The Nocatee Member is named for the town of Nocatee in central DeSoto County, Florida. The type core is W-12050, Hogan #1, located in the SE 1/4, NW 1/4, Sec. 16, T. 38 S., R. 26 E., DeSoto County, Florida, with a surface elevation of 19 m. The type Nocatee occurs between –89.5 m MSL and –15.8 m MSL. The type core was drilled by the Florida Geologic Survey.

THE TAMPA MEMBER, ARCADIA FORMATION

Because of its lithologic differences with, and chronological isolation from, the main Type Arcadia beds (late Chattian–Aquitanian as opposed to mid-Burdigalian–Langhian), the senior author (Petuch, 2004: 16) recently elevated the Tampa to full formational status, retaining the nomenclature of the early workers. In this book, however, we have decided to return to Scott's stratigraphic scheme and here consider the Tampa to be only a member of the Arcadia Formation. Future workers may wish to revisit the lithostratigraphic and nomenclatural status of this distinctive and historically important unit.

History of Discovery

Surface outcrops of the Tampa Member at Tampa Bay have been known for over 100 years and were first studied by Angelo Heilprin during his investigations of the *Okeechobee Wilderness* (1886). Based on Heilprin's research on the "Silex Beds," L.C. Johnson (1883) proposed the name Tampa Formation for these fossiliferous outcrops. Dall (1892) later elevated the Tampa to group status, bringing together the Chipola Formation and "Alum Bluff Beds" (now known to be Chipola outcrops on the Apalachicola River) of the Florida Panhandle and the Tampa "Silex Beds" at Tampa Bay. Finally, Matson and Clapp (1909) formally described the "Tampa Formation" and restricted the use of the name to only the beds exposed in the vicinity of Tampa. More recently, King and Wright (1979) redescribed the Tampa Formation in detail and designated a stratotype (at Ballast Point, Hillsborough County). Later, Scott (1988: 73) reduced the status from *formation* to *member*, stating that " … the reduction is necessary due to the limited areal extent of the Tampa and its interfingering, gradational nature with part of the Arcadia Formation."

Lithologic Description and Areal Extent

While the Nocatee Member was deposited at the mouth of river systems emptying into the Tampa Subsea, the Tampa Member was deposited within the carbonate reef lagoon areas of the Hillsborough Lagoon System, on the coral reefs of the Tampa Reef Tract, and along the coral cays of the Tampa Archipelago (see Petuch, 2004: 70–77). For this reason, carbonates dominate the Tampa Member.

The best lithologic description of the Tampa was given by Scott (1988: 70) and is as follows (condensed here):

> … The Tampa Member consists predominantly of limestone and subordinate dolostone, sands, and clays … Phosphate grains generally are present in the Tampa in amounts less than 3%, although beds containing greater percentages do occur, particularly near the facies change limits of the member … the limestones are variably quartz sandy and clayey with minor or no phosphate. Fossil molds are often present and include mollusks, foraminifera, and algae. Colors range from white to yellowish-gray. The limestones range from mudstones to packstones but are most often wackestones. The dolostones often contain fossil molds, similar to those in the limestones, and are variably quartz sandy and clayey with minor or no phosphate. They are typically microcrystalline to very fine grained and range in color from pinkish-gray to light olive-gray … Siliceous beds are often present in the updip portion of the Tampa. In the type area near Tampa Bay, the unit is well known for its silicified corals, siliceous pseudomorphs of many different fossils, and chert boulders.

Siliceous pseudomorphs of gastropod mollusks have been found by divers in Tampa exposures on the stream bed of the Peace River near Arcadia, DeSoto County.

The Tampa Member is geographically restricted to the west central coast of the Floridian Peninsula, from Hillsborough and Pinellas Counties, through Hardee, Manatee, Sarasota, and DeSoto Counties south to northernmost Charlotte County (Figure 2.8). Throughout its range, the Tampa varies in thickness from 82 m at Sarasota, Sarasota County, to 42 m at Bradenton, Manatee County, with an average thickness of 30.5 m. The top of the Tampa ranges in elevation from +23 m

MSL in northeastern Hillsborough County to –98.5 m MSL in northern Sarasota County to over –100 m MSL in southern Manatee County (probably following the deep areas associated with the western side of the Yeehaw Strait).

Stratotype

From Scott (1988: 68):

> ... King and Wright (1979) thoroughly discussed the Tampa Member (their Tampa Formation) and its type locality. They designated Ballast Point core W-11541, Hillsborough County as the principal reference core (SE $^1/_4$, NW $^1/_4$, Sec. 11, T. 30 S., R. 18 E.). The Tampa Member occurs from –2.7 m MSL to –22.5 m MSL in this core.

Age and Correlative Units

Besides the contemporaneous Floridian formations mentioned earlier under the Nocatee Member, the Tampa Member is exactly correlative with the Haywood Landing Member of the Belgrade Formation of North Carolina, the Old Church Formation of Maryland and Virginia, the lower part of the Parachucla Formation of Georgia, and the Paynes Hammock Formation of Mississippi. The Tampa Member spans the entire time frame of the Tampa Subsea, from the late Chattian Oligocene to the Aquitanian Miocene, and straddles the Oligocene–Miocene boundary.

Tampa Index Fossils

At several localities, such as along Tampa Bay, in the Alafia River, and in the Peace River, beds of beautifully preserved fossil corals and mollusks are typically encountered. Like those of the Suwannee Limestone, the Tampa specimens are agatized (siliceous pseudomorphs), usually being a reddish-orange or deep blue color (depending on the incorporation of either oxidized or reduced iron). The amazing richness of the Tampa fauna was summarized and illustrated by Dall (1915) and later added to by Mansfield (1937). The fossil assemblages of the Tampa Member all correspond to environments within the Hillsborough Lagoon System or along the Tampa Reef Tract. The areas south of Orange Island were too deep to support coral growth and the associated molluscan fauna. This explains the limited areal extent of the typical Tampa Member.

The Tampa reef areas contained a typical Hernandoan molluscan fauna. By Tampa time, however, several Suwannee Hernandoan index genera had become extinct. Some of these included *Cestumcerithium*, *Telescopium*, *Prismacerithium*, *Cypraeorbis*, *Suwannescapha*, and many of the classic Mississippi Vicksburg genera. The Tampa molluscan fauna contains a very high percentage of endemic taxa, underscoring the uniqueness of the carbonate lagoons and coral reefs of Tampa time. Some of the more commonly encountered species include:

Gastropoda
Turbo (Marmarostoma) crenorugatus (Figure 2.9B)
Pyrazisinus campanulatus (Figure 2.9G)
Pyrazisinus cornutus
Eichwaldiella atacta (Figure 2.10H)
Calvertitella tampae (Figure 2.9F)
Doxander liocyclus
Orthaulax pugnax
Ampullinopsis streptostoma
Pachycrommium floridanum
Floradusta ballista (Figure 2.9H and Figure 2.9I)
Loxacypraea tumulus (Figure 2.10C and Figure 2.10D)

Melongena turricula (Figure 2.10A)
Spinifulgur perizonatum
Spinifulgur tampaensis
Spinifulgur stellatum
Turbinella polygonatus
Vasum subcapitellum (Figure 2.9E)
Vasum engonatum
Falsilyria musicina (Figure 2.9D)
Oliva (Strephona) posti (Figure 2.9D)
Gradiconus planiceps (Figure 2.9A)
Gradiconus iliolus
Cerion anodonta
Bivalvia
Anadara latidentata (Figure 2.10G)
Crassostrea rugifera (Figure 2.10F)
Codakia scurra
Pseudomiltha hillsboroensis
Periglypta tarquinia
Venericardia serricosta (Figure 2.10E)
Lirophora ballista (Figure 2.10E)

Along with these endemic taxa, the Tampa molluscan fauna also shares several species with the Haywood Landing Member of the Belgrade Formation of North Carolina. Some of these widespread taxa include:

Gastropoda
Sinum imperforatum
Ecphorosycon tampaensis (Figure 2.9J)
Tritonopsis biconica (Figure 2.10J)
Pugilina quinquespina
Bivalvia
Chama tampaensis
Dinocardium taphyrium
Anomalocardia floridana
Chione spada

In the reefal facies of the Tampa, a large fauna of corals can be used to delineate the member boundaries. These agatized corals are collectively designated as the Florida State Rock (Tampa Bay agatized coral). Some of the more commonly encountered species include:

Cnidaria–Scleractinia
Porites anguillensis (Figure 2.10I)
Acropora saludensis subspecies
Goniopora canalis
Montastrea tampaensis (Figure 2.9C)
Montastrea silicensis
Siderastrea silicensis
Siderastrea hillsboroensis
Antiguastrea cellulosa (Figure 2.10K)
Meandrina dumblei

FIGURE 2.9 Index fossils for the Tampa Member of the Arcadia Formation, Hawthorn Group. A = *Gradiconus planiceps* (Heilprin, 1886), length 49 mm; B = *Turbo (Marmarostoma) crenorugatus* Heilprin, 1886, length 28 mm; C = *Montastrea tampaensis* Vaughan, 1919, length 87 mm; D = *Falsilyria musicina* (Heilprin, 1886), length 41 mm; E = *Vasum subcapitellum* Heilprin, 1886, length 36 mm; F = *Calvertitella tampae* (Heilprin, 1886), length 64 mm; G = *Pyrazisinus campanulatus* (Heilprin, 1886), length 41 mm; H, I = *Floradusta ballista* (Dall, 1915), length 23 mm; J = *Ecphorosycon tampaensis* (Dall, 1892), length 21 mm; K = *Vericardia serricosta* (Heilprin, 1886), length 28 mm. All specimens are siliceous pseudomorphs, replaced by agate.

FIGURE 2.10 Index fossils for the Tampa Member of the Arcadia Formation, Hawthorn Group. A = *Melongena turricula* Dall, 1890, length 90 mm; B = *Oliva (Strephona) posti* Dall, 1915, length 23 mm; C, D =

FIGURE 2.10 (continued) *Loxacypraea tumulus* (Heilprin, 1886), length 25 mm; E = *Lirophora ballista* (Dall, 1903), length 22 mm; F = *Crassostrea rugifera* (Dall, 1898), length 42 mm; G = *Anadara latidentata* Dall, 1898, length 30 mm; H = *Eichwaldiella atacta* (Dall, 1915), length 31 mm; I = *Porites anguillensis* Vaughan, 1919, length 158 mm; J = *Tritonopsis biconica* (Dall, 1915), length 27 mm; K = *Antiguastrea cellulosa* (Duncan, 1863), length 71 mm. All specimens are siliceous pseudomorphs, replaced by agate.

The marine communities of the Tampa Member (Tampa Reef Tract and Hillsborough Lagoon areas) were described by the senior author (Petuch, 2004: 70–77) and include: the *Cerion anodonta* Community (hypersaline pools, shore grass environment); the *Pyrazisinus campanulatus* Community (mud flat and mangrove forest environment); the *Periglypta tarquinia* Community (Turtle Grass bed environment); the *Porites anguillensis* Community (back reef environment); and the *Montastrea tampaensis* Community (reef platform environment). The demise of the Tampa molluscan fauna marked the end of the Hernandoan Mollusk Age.

PALEOGEOGRAPHY OF THE ARCADIA SUBSEA, OKEECHOBEAN SEA

At the end of Aquitanian time, sea levels again dropped precipitously, and the Okeechobean Sea area was again emergent. For over 4 million years, terrestrial conditions prevailed, and river systems on the highlands that had been Orange Island continued to deposit voluminous amounts of sediments into the Everglades area. During this time, the Tampa Reef Tract, Tampa Archipelago, and Hillsborough Lagoon System were covered by fluvial deposits (some of the quartz sand stringers seen in the lower Type Arcadia), permanently altering the coastal environments. Never again did zonated coral reef tracts grow that far north on the Floridian Peninsula. Fluvial sediments from southern Georgia also poured into the Suwannee Strait area, connecting Florida with the Georgia mainland. For the first time, Florida had become a permanent peninsula. Southward, within the Everglades area, large freshwater lakes probably occupied the depressional feature of the Everglades Unconformity.

By the early middle Burdigalian Miocene, climates became warmer and sea levels rose to their previous heights. The center of coral reef building had now shifted southward to a new Okeechobean subsea, the Arcadia Subsea (named for Arcadia, DeSoto County) (Figure 2.11). Here, the Collier Banks had become greatly enlarged, being fed by siliciclastics and eroded carbonate sediments from the southern end of the Floridian Peninsula. These banks supported extensive oyster bars and scattered coral bioherms. The eastern Dade Banks remained narrow and undeveloped, not having received the large sediment budget from the southwestern edge of the peninsula. The deep Everglades Basin area was rapidly filling with a mixture of siliciclastics, carbonate fines, and plankton resources to produce the sediments of the Type Arcadia Formation (described earlier in this chapter). The western side of the Arcadia Subsea, east of the Collier Banks, was shallower than the eastern side adjacent to the Dade Banks. This deep feature, the **Loxahatchee Trough** (named for Loxahatchee, Palm Beach County), is readily seen in the plunging contours of the Arcadia Formation surface shown by Scott (1988; Figure 63 and Figure 64). The areal distribution of the Type Arcadia Formation is shown on Figure 2.12.

In the north, the Suwannee Strait was now closed, filled in by a narrow isthmus of mangrove forests growing on quartz sand and clays. To the west, within the Chipola Subsea of the Choctaw Sea (see Petuch, 2004), the **Torreya Lagoon System** (named for the Torreya Formation) contained extensive oyster beds (the ostreid *Ostrea* and the anomiid *Carolia*) that were associated with the mangrove forests. On the eastern side of the narrow isthmus, a long, narrow bay, the **Parachucla Embayment** (named for the Parachucla Formation of northeastern Florida and Georgia) was all that remained of the original Suwannee Strait. By the late Miocene, the Parachucla Embayment was filled by terriginous sediments, greatly increasing the width of the Floridian Peninsula. After that time, sediments transported by longshore currents moved directly down the Florida coast, forming the first extensive barrier islands and coastal lagoons along the eastern shore.

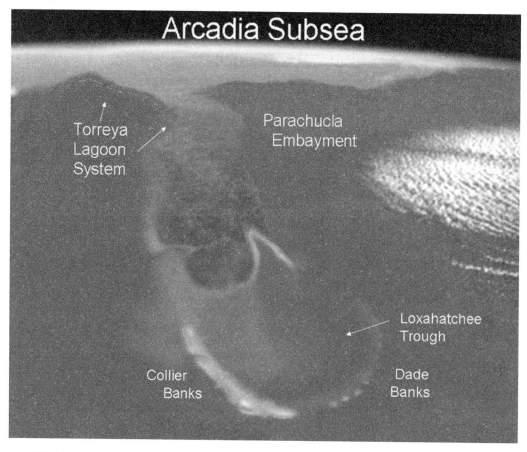

FIGURE 2.11 Simulated space shuttle image of the Floridian Peninsula (altitude: 100 mi) during the Burdigalian Miocene, showing the possible appearance of the Arcadia Subsea of the Okeechobean Sea and some of its principal geomorphologic features.

The upwelling systems that had become established during the Tampa and Nocatee times were now stronger and more extensive, and were probably permanent features along the western coast of the peninsula and the Collier Banks. These upwellings may have been caused by the formation of the Apalachee Gyre, a cool-water gyre off the Gulf Loop Current (Petuch, 2004: 157–159). With the closing of the Suwannee Strait, the Gulf of Mexico current structure was reorganized, with an accelerated flow across the shallow western continental shelf. This abutment would have entrained deeper, nutrient-rich water up into the neritic zone, producing immense, probably permanent, phytoplankton blooms. These, in turn, were carried into the Okeechobean Sea basin, contributing to the phosphatic component of the Type Arcadia Formation and supporting accelerated biogenic deposition on the Collier Banks. The Apalachee Gyre continued into the early Piacenzian Pliocene only in the northern Gulf, producing the cool water conditions in the Jackson Subsea of the Choctaw Sea (lower Jackson Bluff Formation; see Petuch, 2004: 157–162).

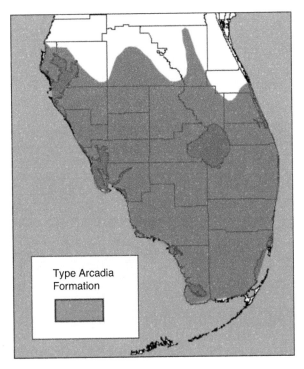

FIGURE 2.12 Areal distribution of the Type Arcadia Formation, Hawthorn Group.

View of the Everglades Agricultural Area (EAA), showing the vast sod and sugar cane fields south of Lake Okeechobee in western Palm Beach County. The area shown here is typical of the Loxahatchee Trough region, between the Big Cypress Spur and the Atlantic Coastal Ridge. The lake in the foreground is one of the flooded fill pits in the Griffin Brothers quarries and sod farm. At this locality, the Hawthorn Group lies almost 100 meters below the surface.

3 Middle and Late Miocene to Early Pliocene Southern Florida

The middle Miocene (Langhian and Serravallian Ages), the late Miocene (Tortonian and Messinian Ages), and the early Pliocene (Zanclean Age) were times of chaotic climatic conditions, with long, mild spells separated by severe cold periods (Petuch, 2004). During the late Langhian, and also during the exceptionally severe cold time and sea level drop during the early Serravallian, the entire Everglades area was emergent. Only in the late Serravallian and early Tortonian did the Okeechobean Sea basin begin to refill, resulting in the Polk Subsea and, further into the late Tortonian and early Messinian, in the subsequent Charlotte Subsea. After the late Messinian regression (shown on Figure 4.2, Chapter 4), this pattern was also repeated in the early Zanclean, by the reflooding of the Everglades area with the Murdock Subsea. Simulated space shuttle images of these three subseas are illustrated in color in Chapter 1. The late Miocene and early Pliocene were also the times of the great deltas (Cunningham et al., 2003), when the Okeechobean Sea basin was rapidly filled with the sediments of two giant deltaic complexes (the DeSoto and Immokalee Deltas; discussed later in this chapter).

PALEOGEOGRAPHY OF THE POLK SUBSEA, OKEECHOBEAN SEA

Throughout most of the Serravallian Miocene, during the great mid-Miocene cold time and sea level drop (Cunningham et al., 2003; Petuch, 2004: 261, and the corresponding "Transmarian Extinction"), the Okeechobean Sea basin remained emergent. At this time, the Everglades area was a shallow depression, probably containing one or more large freshwater lakes. In the late Serravallian, the marine climate rapidly warmed and sea levels rose quickly, flooding the Everglades Basin with the Polk Subsea. Existing prior to the formation of the deltas, the Polk Subsea was a deep, essentially circular lagoon bounded by a series of narrow, shallow banks (Figure 3.1). In the northwest, a series of broad, shallow lagoons, the **Polk Lagoon System** (named for Polk County), comprised most of the coastal area of peninsular southern Florida. As in the older Tampa and Arcadia Subseas, permanent upwelling systems occurred just offshore, flooding the Polk Lagoon System and main Okeechobean basin with high-productivity, plankton-rich water. Accumulating on the floor of the Polk Lagoon System, these plankton oozes became the primary source of the phosphates and phosphorite beds of the Peace River Formation.

Along the western side of the Okeechobean basin, the chain of low, shallow oyster-covered banks that had formed during the Dade, Tampa, and Arcadia Subseas time, the **Collier Banks** expanded to become a larger and more prominent feature. Feeding on the rich phytoplankton resources, these oyster bioherms grew to massive proportions. The **Dade Banks,** which had also formed in the older Dade, Tampa, and Arcadia Subseas, and the Arcadia Subsea-derived **Loxahatchee Trough,** remained essentially unchanged, being exposed to more open oceanic conditions. Presaging the deltas of later times, large amounts of clay minerals were pouring into the Okeechobean basin at this time, being discharged from rivers draining into the Polk Lagoon System. These clays and muds, mixed with plankton-derived phosphates, produced the Everglades component of the Peace River Formation.

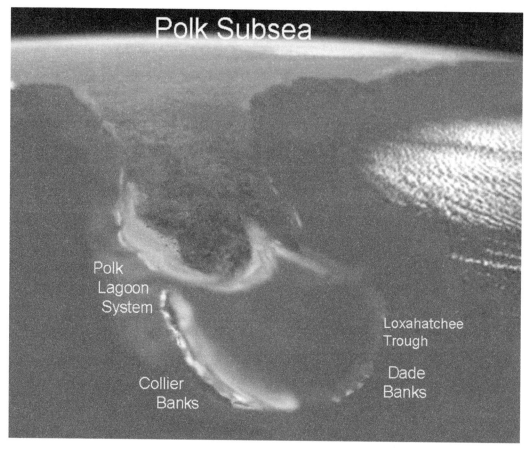

FIGURE 3.1 Simulated space shuttle image (altitude: 100 mi) of the Floridian Peninsula during the late Serravallian Miocene, showing the possible appearance of the Polk Subsea of the Okeechobean Sea and some of the principal geomorphologic features.

THE PEACE RIVER FORMATION, HAWTHORN GROUP

THE BONE VALLEY MEMBER, PEACE RIVER FORMATION

History of Discovery

Although known for almost a century as a source of phosphate fertilizers (as the Hawthorn Formation or Hawthorn clays, informally named by Dall, 1892), the Peace River Formation was not formally described or defined until 1988, when Scott published his monograph on the Hawthorn Group. From this thick sequence of sediments and distinct phosphorite beds, he recognized a type — Peace River, which makes up most of the formational sequence, and an upper Bone Valley Member, which is essentially nonexistent in the Everglades area. The Bone Valley Member, which is prevalent in northern and central Florida, was first recognized by Matson and Clapp in 1909, but was referred to only as the "Bone Valley Gravel." Scott's Peace River, then, is a combination of the older and poorly defined "Hawthorn Formation" and "Bone Valley Formation."

Scott (1988: 79) also included the Bayshore and Murdock Station Formations of Hunter (1968, originally described as members of the Tamiami Formation) in his Peace River Formation, as the uppermost components (also referred to as the "Upper Peace River" by Cunningham et al., 2003). Subsequent studies within the Everglades area, however, have shown that the areal extents, thicknesses, and developments of these formations were larger and more complicated than previously

FIGURE 3.2 Areal distribution of the Peace River Formation (undifferentiated), Hawthorn Group.

thought. As the Bayshore and Murdock Station are so closely tied to, and essentially define, the DeSoto and Immokalee Deltas (discussed later in this chapter), we feel it is more expedient to consider them as full, separate formations and recommend that they be removed from the Peace River (as the Upper Peace River). Because of the removal of the late Tortonian and Messinian Bayshore and the Zanclean Pliocene Murdock Station components, the Peace River Formation now is chronologically more restricted than originally thought, dating only from the late Serravallian and early Tortonian Miocene, and not the Messinian Miocene and Zanclean Pliocene, as originally proposed. As pointed out by Scott (1988: 86), the Peace River below the Everglades region is undifferentiated, with well-defined Bone Valley sediments occurring only rarely and sporadically.

Lithologic Description and Areal Extent

As described by Scott (1988: 79) (condensed here):

> … The Peace River Formation consists of interbedded quartz sand, clays, and carbonates. The silici-clastic component predominates and is the distinguishing lithologic feature of the unit. Typically, the siliciclastics comprise two thirds of the formation. The quartz sands are characteristically clayey, calcareous to dolomitic, phosphatic, very fine to medium grained, and poorly consolidated. Their color ranges from light gray and yellowish-gray to olive-gray. The phosphatic content of the sands is highly variable. In the type section, the phosphate content is lowest in the upper part of the section and greatest near the base … The phosphate occurs as sand and gravel-sized particles. The gravels are most abundant in the Bone Valley Member, although they may occur elsewhere in the unit … Clay beds are quite common in the Peace River Formation … the clay minerals consisting of smectite (montmorillonite), palygorskite (attapulgite), and sepiolite … smectite and palygorskite are the dominant clay minerals in the formation … Carbonates occur throughout the Peace River Formation. Characteristically, they comprise less than 33% of the Peace River section. The carbonates may be either limestone or dolostone. Updip (northward), dolostone occurs more frequently. The limestones are variably sandy, clayey, and phosphatic, poorly indurated, mudstones to wackestones. They vary in color from yellowish-gray to white … Mollusk molds are common throughout the carbonates.

The Peace River Formation underlies the entire Everglades region (Figure 3.2). Here, it occurs at depths of approximately 40–46 m below mean surface. The thickness of the Peace River within the Everglades Basin reflects the bathymetry of the Polk Subsea. In the west, along the shallower areas near the Collier Banks, the formation has thicknesses that average 100–122 m. In the east, within the deepwater area of the Loxahatchee Trough, thicknesses may vary between 198–200 m. Unaware of the existence of this deep paleotrough feature at the time, Scott (1988: 81) remarked that the Peace River sequence in Palm Beach County seemed "anomalously thick." Farther north, along the northern shoreline of the Polk Subsea, the Peace River occurs closer to the surface, often only at 10–20 m depths.

Stratotype

From Scott (1988: 79): … "The type section of the Peace River Formation is designated as core W-12050, Hogan #1, located in east central DeSoto County, Florida (SE $^{1}/_{4}$, NW $^{1}/_{4}$, Sec. 14, T. 38 S., R. 23 E.), with a surface elevation of 62 feet (19 m)."

Age and Correlative Units

Based on the presence of Hemphillian mammal fossils (Scott, 1988: 84) and on Maryland Miocene-type bivalves (St. Mary's Formation, recognizable molds and casts), the Peace River Formation is now known to span the period from the late Serravallian to the early Tortonian Miocene, a range of over 4 million years. The formation is exactly correlative with the St. Mary's Formation, Chesapeake Group of Maryland (including the late Serravallian Conoy and Little Cove Point Members, and the early Tortonian Windmill Point and Chancellor's Point Members) and with the Shoal River Formation of the Alum Bluff Group of the Florida Panhandle (Petuch, 2004).

Peace River Index Fossils

The Peace River sediments in the Everglades area were heavily leached by infiltrating groundwater from the subsequent DeSoto and Immokalee Deltas, particularly during emergent times. Because of this infiltration and dissolution, molluscan fossils are present only as casts and molds. Some of the more easily recognizable include:

Gastropoda
Torcula subvariabilis
"*Akleistostoma*" species (Figure 3.3E and F) (may be a Muracypraea)
Calusacypraea species (Figure 3.3B and C)
Gradiconus deluvianus
Oliva (Strephona) species
Bivalvia
Perna incurvus (Figure 3.3G)
Clementia inoceriformis (Figure 3.3A)
Cyrtopleura arcuata (Figure 3.3D)

A single marine community from the Peace River Formation was described by the senior author (Petuch, 2004: 130), the *Clementia inoceriformis* Community (shallow muddy sand lagoon and Turtle Grass beds).

FIGURE 3.3 Index fossils for the Peace River Formation, Hawthorn Group. A = *Clementia inoceriformis* (Wagner, 1839), length 68 mm; B,C = *Calusacypraea* new species, length 46 mm (oldest-known *Calusacypraea*); D = *Cyrtopleura arcuata* (Conrad, 1841), length 122 mm; E,F = "*Akleistostoma*" new species, length 48 mm (by having distinct dorsal tubercles, this species may be a Floridian representative of the genus *Muracypraea*); G = *Perna incurvus* (Conrad, 1839), length 120 mm. All specimens are casts from moldic dolomite.

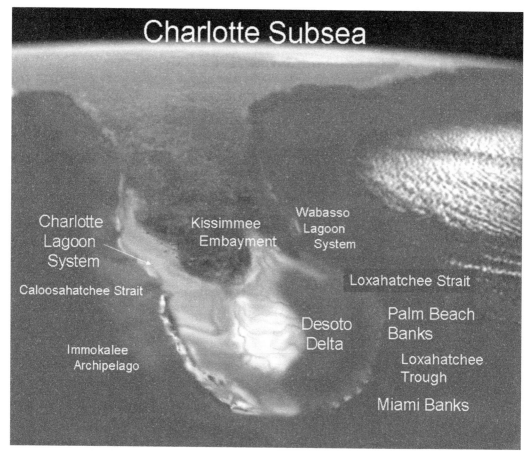

FIGURE 3.4 Simulated space shuttle image (altitude: 100 mi) of the Floridian Peninsula during the early Messinian Miocene, showing the possible appearance of the Charlotte Subsea and some of its principal geomorphologic features.

PALEOGEOGRAPHY OF THE CHARLOTTE SUBSEA, OKEECHOBEAN SEA

During the late Miocene (Tortonian to early Messinian), a series of warm climatic intervals led to higher sea levels and a reflooding of the Okeechobean Sea basin. This new marine system, the Charlotte Subsea, was geomorphologically complex and represented an enlarged and infilled version of the older Polk Subsea (Figure 3.4). In the intervening emergent time, between the Polk and Charlotte Subseas (mid-to-late Tortonian sea level drop), a large deltaic complex, the **DeSoto Delta** (named for DeSoto County), began to feed into the Okeechobean Sea basin from a paleovalley, in what is now the DeSoto Plain and Lake Wales Ridge region (Cunningham et al., 2003). This huge clastic wedge spread to fill the entire western half of the Okeechobean basin and fed into a freshwater lake that existed along the southern edge of the Everglades. Infiltrating groundwater from this long-lived lake and delta complex leached and dissolved away all the aragonitic and calcitic fossils that occurred in the underlying Peace River sediments. In the late Tortonian, the DeSoto Delta was flooded by the rising sea levels of the Charlotte Subsea. This drowned delta formed a sedimentary platform upon which unique Charlotte marine ecosystems became established (Petuch, 2004: 130).

The northern edge of the Charlotte Subsea was bordered by three lagoon systems: the **Charlotte Lagoon System** (named for Charlotte County) in the west, the **Kissimmee Embayment** (named for the Kissimmee River) in the center, and the **Wabasso Lagoon System** (named for Wabasso, Indian River County) in the east. The western edge of the DeSoto Delta was bordered by the Immokalee Archipelago, a chain of low sand islands and oyster bars (bioherms) that extended from present-day Port Charlotte southward to Naples. This archipelago prevented the spreading of the deltaic sediments westward into the Gulf of Mexico, in turn, funneling this clastic wedge into the main basin of the Okeechobean Sea.

The eastern half of the Charlotte Subsea basin still contained a deep water (over 200 m) lagoon, the **Loxahatchee Trough**, and this depression, which resembled a miniature Tongue-of-the-Ocean (Bahamas), was bordered on the east by a narrow, curving chain of oyster and coral bioherms, the **Palm Beach Banks** and the **Miami Banks**. The nearly enclosed Charlotte Subsea was connected to the Gulf of Mexico by the broad, shallow **Caloosahatchee Strait** (named for the Caloosahatchee River) in the west and by the deep **Loxahatchee Strait** (named for the Loxahatchee River) in the east. Once established in the Tortonian and Messinian, the Caloosahatchee Strait, Kissimmee Embayment, Loxahatchee Strait, and Loxahatchee Trough continued to persist in all the subsequent subseas until the end of the Pleistocene. Of special interest during this time was the existence of extensive and long-lived upwelling systems along the western coast of Florida. These cool and nutrient-rich water masses, which were probably produced by a strong, southward-flowing branch of the Apalachee Gyre (Petuch, 2004: 157), supported immense phytoplankton blooms all along the Charlotte Lagoon System and the Immokalee Archipelago. These high-productivity waters and their plankton biomass were the principal sources for the phosphorites and phosphate grains that are so prominent in the sediments of the Bayshore Formation.

THE BAYSHORE FORMATION, HAWTHORN GROUP

HISTORY OF DISCOVERY

Until recognized as a set of distinctive, phosphate-bearing beds by Hunter in 1968, this formation was generally overlooked by most previous workers or simply considered to be part of the older Miocene Bone Valley and Peace River beds. Originally, Hunter included her new unit in the Tamiami Formation, Okeechobee Group, considering it to be the lowest and oldest member. As discussed previously, Scott (1988) later placed Hunter's Bayshore "Member" and Murdock Station "Member" (discussed later in this chapter) in his Peace River Formation, Hawthorn Group. Because of the presence of well-preserved mollusks and barnacles, which constitute a large carbonate fraction, the Bayshore can be differentiated from the underlying Peace River and can be seen to constitute a separate geological unit. The formation was recently informally removed from member status in the carbonate-rich Tamiami Formation, Okeechobee Group (as proposed by Hunter), and was elevated to full formational status in the phosphate-rich Hawthorn Group (Petuch, 2004: 18).

LITHOLOGIC DESCRIPTION AND AREAL EXTENT

Overall, the Bayshore Formation is lithologically distinctive and is easily separated from the underlying Peace River Formation and the overlying Murdock Station Formation. Typically, the Bayshore is composed of clays (smectite and palygorskite) and sandy clays, usually white, tan, or pale orange in color, and contains variable amounts of black phosphorite grains and phosphate-coated pebbles. Some facies, such as those encountered in canal digs near Murdock, Charlotte County, are sandier, light gray in color, and contain higher amounts of phosphates. Characteristically, thick beds of well-preserved mollusk shells (typically scallops, oysters, and ecphoras) and barnacles (Cirripedia) are present and make up a large percent of the carbonate fraction. The formation was

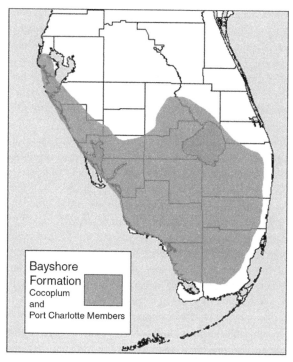

FIGURE 3.5 Areal Distribution of the Bayshore Formation, Hawthorn Group, and its two members, the Cocoplum and the Port Charlotte.

found to be better developed and more widely distributed than was originally thought by Hunter, and has been shown to consist of two separate members, namely, the lower **Cocoplum Member** (Tortonian) and the upper **Port Charlotte Member** (Messinian) (both informally proposed here as new members and described in the following sections). The Cocoplum Member, which is separated by a large unconformity from the overlying Port Charlotte Member, also contains sandy dolomite beds that contain large amounts of phosphorites.

These beds vary in color from light orange and buff to pale gray. (Hunter, original description: " ... white to light tan in color; sandy to very sandy clay containing black and brown phosphate grains and small pebbles. Well-preserved calcite fossils are present ...").

The Bayshore Formation, including both the Cocoplum and Port Charlotte Members, underlies most of the Everglades region (Figure 3.5). The thickest and best-developed areas of the Bayshore are known primarily from the Port Charlotte and southern Sarasota County areas. Here, the formation is estimated to be over 2 m thick and in some areas, such as along the Cocoplum Canal, southern Sarasota County, the formation may reach thicknesses of over 3 m. In all these localities, the Bayshore occurs at depths ranging from 4 to 7 m below mean surface. Based on the distribution of key molluscan index fossils, the formation is now known to extend from Pinellas County south to northern Monroe County, and from Okeechobee and Highlands Counties in the Kissimmee Valley south to Dade County. Within the Okeechobee Plain areas, Bayshore index fossils have been encountered in deep dredging (approximately 20 m) by draglines during Kissimmee River channel deepening.

The smectite and palygorskite (brackish water) clays of the Bayshore Formation are distinctly deltaic and reflect the depositional influence of the DeSoto Delta throughout the Tortonian and Messinian. Subsequent groundwater infiltration from the Immokalee Delta during the Zanclean Pliocene led to further leaching and diagenesis. This deltaic leaching also dissolved all Bayshore aragonitic fossils, leaving only calcitic species.

STRATOTYPE

From Hunter (1968: 443): " … Canal bank, Section 3, T. 40 S., R. 21 E., on Bayshore Waterway, $1/2$ mile south of U.S. 41, Port Charlotte, Charlotte County, Florida." The formation was described from an underwater exposure, from near the surface to approximately 1 m below surface.

A classic stratigraphic column for the Bayshore sequence was given by Hunter (1968: 443) and is reproduced as follows (with modifications by ourselves in light of new data):

Bed	Description	Thickness (in feet)
	Pleistocene Series	
	Fort Thompson Formation	
6	Sand	4.0
5	Shelly sand with *Pyrazisinus gravesae* (originally referred to as *P. scalatus*)	2.0
4	Sand	1.5
	Pliocene Series (Originally Thought to Be Miocene)	
	Murdock Station Formation (originally placed in the Tamiami Formation)	
3	Pecten biostrome, dark gray pecten shells and a few barnacle fragments (lower bed of Murdock Station)	4.5
2	Gray phosphatic sand to and below water level	2.5+
	Miocene Series	
	Bayshore Formation (Port Charlotte Member)	
1	Sandy clay from below water level containing *Globecphora floridana* (originally referred to as *Ecphora quadricostata*) and *Chesapecten middlesexensis* (originally referred to as *Pecten santamaria middlesexensis*)	1.0+

AGE AND CORRELATIVE UNITS

Based upon the presence of key index fossils, such as *Ecphora whiteoakensis*, *Chesapecten middlesexensis*, and *Carolinapecten urbannaensis*, the Bayshore Formation is now known to date from the latest Miocene, from the late Tortonian and Messinian. The Bayshore is exactly correlative with the Eastover Formation, Chesapeake Group of Virginia and northern North Carolina. The two members of the Bayshore Formation, the Cocoplum and Port Charlotte, are separated by an unconformity that represents almost 1 million years of missing time.

THE COCOPLUM MEMBER, BAYSHORE FORMATION

Lithologic Description and Areal Extent

This new member characteristically contains large amounts of sandy dolosilts and thick sandy dolomite stringers, most often of an orange or a tan-buff color, but sometimes white, and containing variable amounts of black phosphorite granules. Clays, most commonly palygorskite and smectite, are also present in variable amounts and are colored a light tan or orange-tan. Molluscan bioclasts are abundant in this member, consisting mostly of oysters and muricid gastropods, and make up the largest part of the carbonate fraction. Gastropod fossils from this member are colored a distinctive golden yellow or pale orange. The uppermost part of the Cocoplum Member indurates into a semifriable, grayish sandy dolomite/arenaceous calcarenite that is rich in phosphate grains. Oysters from this indurated unit (Figure 3.6) are usually colored a gray-tan. The Cocoplum Member has the same areal distribution as the entire Bayshore Formation, extending from Pinellas County to Monroe and Dade Counties. The classic Cocoplum index fossils *Ostrea brucei* and *Ecphora whiteoakensis* have been collected during dredgings in the Kissimmee River (20 m depth) near the Osceola–Okeechobee County line. Throughout its whole range, the Cocoplum Member has thicknesses that average around 1.5 m but may be thicker in the northern Kissimmee River Valley.

Stratotype

The type section of the Cocoplum Member is found along the east side of the Cocoplum Waterway Bridge at North Port, Sarasota County, on the Cocoplum Waterway just north of the Sarasota County–Charlotte County line. The type sedimentary material and index fossils are exposed at times of lowest water level, just above the water surface (Hunter, 1968: 443). The new member is named for the Cocoplum Waterway, Sarasota County, the type locality. Cocoplum sediments and index fossils were collected along 2 km of the canal bank.

Age and Correlative Units

The Cocoplum Member is now known to date from the late Tortonian Miocene and is exactly correlative with the Claremont Manor Member of the Eastover Formation, Chesapeake Group of Virginia and northern North Carolina. Both the Claremont Manor and the Cocoplum Members share the chronologically sharply defined mollusks *Ostrea brucei* and *Ecphora whiteoakensis*.

Cocoplum Index Fossils

As the Cocoplum sediments are highly leached, only thick-shelled calcitic fossils are preserved. Some of the more abundant and indicative include:

Gastropoda
Ecphora whiteoakensis (Figures 3.6A and B) (*E. hopei* Petuch, 1991, originally incorrectly
   thought to have come from the younger Tamiami Formation, is a synonym)
Zulloia violetae (Figures 3.6D and E)
Bivalvia
Gigantostrea leeana (Figures 3.6F and G)
Ostrea brucei (Figure 3.6C)
Ostrea disparilis subspecies
Cirripedia
Chesaconcavus species

A single Cocoplum marine community was described by the senior author (Petuch, 2004: 132), the *Gigantostrea leeana* Community (shallow sand-bottom lagoon and oyster bars).

THE PORT CHARLOTTE MEMBER, BAYSHORE FORMATION

Lithologic Description and Areal Extent

This new member characteristically is composed of sandy clay (containing smectite, palygorskite, and sepiolite) that is colored white, yellow, or tan, with large amounts of black phosphate grains and phosphate-coated pebbles. Some facies contain large amounts of highly phosphatic quartz sand, with very little clay, and are light gray or tannish-gray in color. Typically, the Port Charlotte contains abundant well-preserved molluscan bioclasts, predominantly large scallops, oysters, and muricid gastropods, which make up the greatest part of the carbonate fraction. Dolosilts are occasionally present, but make up only a minor fraction of some facies. The Port Charlotte Member has the same areal distribution as the entire formation, extending from Pinellas County to Monroe and Dade Counties and into the Kissimmee Valley of the Okeechobee Plain. Throughout its range, the Port Charlotte has thicknesses that average around 1–2 m.

FIGURE 3.6 Index fossils for the Cocoplum Member of the Bayshore Formation. A,B = *Ecphora whiteoak-ensis* Ward and Gilinsky, 1988, length 101 mm; C = *Ostrea brucei* Ward, 1992, length 97 mm; D = *Zulloia violetae* Petuch, 1994 (variant with rounded varices), length 33 mm; E = *Zulloia violetae* Petuch, 1994 (typical form), length 34 mm; F,G = *Gigantostrea leeana* (Wilson, 1987), length 138 mm.

Stratotype

The type section of the new member is the canal bank on Bayshore Drive, along the Bayshore Canal, Port Charlotte, Charlotte County, Florida (Sec. 3, T. 40 S., R. 21 E., Charlotte County). Here, the Port Charlotte Member is exposed at water level (during low-water times) in the Bayshore Canal, approximately 4 m below land surface, and has a mean thickness of 1.5–2 m. The new member is named for Port Charlotte, Charlotte County, Florida.

Age and Correlative Units

The Port Charlotte Member is now known to date from the early Messinian Miocene and is exactly correlative with the Cobham Bay Member of the Eastover Formation, Chesapeake Group of Virginia and northern North Carolina. Both the Port Charlotte and Cobham Bay Members share the chrono-logically sharply defined mollusks *Chesapecten middlesexensis*, *Carolinapecten urbannaensis*, and *Ostrea geraldjohnsoni*, all classic guide fossils for the early Messinian Miocene of the Atlantic Coastal Plain.

Port Charlotte Index Fossils

Like the Cocoplum Member, the Port Charlotte is highly leached and contains only thick, resistant calcitic fossils. Some of the more indicative and abundant include:

Gastropoda
Globecphora floridana (Figure 3.7A and B)
Globecphora floridana streami (Figure 3.7F) (confined to the uppermost beds of the Port Charlotte; incorrectly listed as coming from the younger Murdock Station Formation by Petuch, 2004: plate 43, and the Murdock Station species should be listed as *Ecphora roxaneae*)
Calusathais handgenae (Figure 3.7C)
Zulloia zulloi (Plate 3.7H)
Bivalvia
Carolinapecten urbannaensis (Figure 3.7E)
Chesapecten middlesexensis (Figure 3.7D)
Ostrea geraldjohnsoni (Figure 3.7G)
Ostrea disparilis subspecies
Cirripedia
Chesaconcavus species

A single marine community from the Port Charlotte Member was described by the senior author (Petuch, 2004: 132), the *Chesapecten middlesexensis* Community (deep sand-bottom lagoon scallop beds).

THE PALEOGEOGRAPHY OF THE MURDOCK SUBSEA, OKEECHOBEAN SEA

The late Messinian Miocene was a time of rapidly repeating severe cold times and major sea level drops, and this led to the extinction of many classic Bayshore mollusks, such as *Zulloia*, *Calusathais*, and *Gigantostrea*. Subsequently, the Okeechobean Sea basin remained emergent for over 1 million years, from approximately 5.5 million to 4.5 million years B.P. Marine climates did not become warmer until the early Zanclean Pliocene (4.5 million years). At that time, sea levels again began to rise, and the Everglades area was reflooded as the Murdock Subsea (Figure 3.8). During the intervening emergent time in the late Messinian, the southwestern part of the Okeechobean basin was covered by

FIGURE 3.7 Index fossils for the Port Charlotte Member of the Bayshore Formation. A,B = *Globecphora floridana* (Petuch, 1989), length 110 mm; C = *Calusathais handgenae* (Portell and E.Vokes, 1992), length 58 mm; D = *Chesapecten middlesexensis* (Mansfield, 1936), length 120 mm; E = *Carolinapecten urbannaensis* (Mansfield, 1929), length 115 mm; F = *Globecphora floridana streami* (Petuch, 1994), length 112 mm; G = *Ostrea geraldjohnsoni* Ward, 1992, length 94 mm; H = *Zulloia zulloi* Petuch, 1994, length 36 mm.

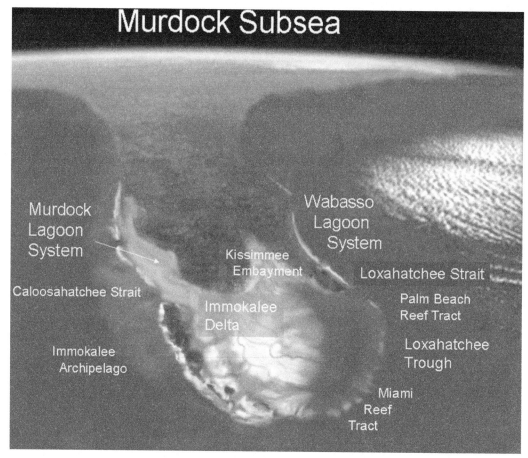

FIGURE 3.8 Simulated space shuttle image (altitude:100 mi) of the Floridian Peninsula during the Zanclean Pliocene, showing the possible appearance of the Murdock Subsea and some of its principal geomorphologic features.

a second large deltaic system, the **Immokalee Delta** (named for Immokalee, Collier County). This prominent geomorphological feature prograded over the older late Miocene DeSoto Delta, filling all deeper depressions along the western side of the Everglades with terriginous sediments. During the reflooding of the Okeechobean basin in the Zanclean Pliocene, the Immokalee Delta was drowned and produced a shallow platform that encompassed the western two thirds of the Murdock Subsea.

Several persistent late-Miocene geomorphological features became larger and more prominent during the early Zanclean sea level rise. The **Immokalee Archipelago** became wider at this time, expanding eastward across the platform formed by the Immokalee Delta. The Palm Beach and Miami Banks began to shoal also, supporting large, shallow-water oyster and coral bioherms. These enlarged Charlotte Subsea features, which form the core of the southern section of the Atlantic Coastal Ridge, became the **Palm Beach Reef Tract** and the **Miami Reef Tract** of the Murdock Subsea. The Wabasso Lagoon System, Loxahatchee Strait, Loxahatchee Trough, Kissimmee Embayment, and Caloosahatchee Strait remained essentially the same as they were during the Tortonian and Messinian Miocene.

As in the older Charlotte Subsea, large upwellings of cold, nutrient-rich water occurred along the western side of the Murdock Subsea. Besides producing colder water conditions, these upwellings flooded the shallow **Murdock Lagoon System** and the Immokalee Archipelago with plankton-rich water, leading to the production of phosphorites. These high-productivity waters supported rich and highly endemic marine communities, comprising immense beds of giant *Chesapecten* scallops, extensive giant barnacle (Cirripedia) "reefs," massive oyster bars, and a rich fauna of marine mammals, including walruses, seals, and whales. The principal depositional unit of the Murdock Subsea, the Murdock Station Formation, represents the last vestige of phosphate deposition and marks the top of the Hawthorn Group.

THE MURDOCK STATION FORMATION, HAWTHORN GROUP

HISTORY OF DISCOVERY

Like the underlying Bayshore Formation, this last phosphate-bearing set of beds was generally overlooked by previous authors until recognized and described by Hunter (1968). Despite the presence of phosphorites, Hunter included this poorly studied unit within her expanded Tamiami Formation. Originally considered to be the second-lowest member of the Tamiami Formation, Hunter named this unit the "Murdock Station Member," with the stratotype near Murdock, Charlotte County (Murdock Station was the old railroad station for Port Charlotte). Like the Bayshore Formation discussed previously, the Murdock Station Formation was placed within the Peace River Formation by Scott (1988) and later considered part of the "Upper Peace River" (Cunningham et al., 2003). Because of its unique lithology and larger development than what was originally thought, the Murdock Station was given full formational status by the senior author (Petuch, 2004: 18).

LITHOLOGIC DESCRIPTION AND AREAL EXTENT

This uppermost component of the Hawthorn Group is lithologically distinct across the entire extent and is easily differentiated from the overlying Tamiami Formation and the underlying Bayshore Formation. As described by Hunter (1968), the type Murdock Station is composed of coarse beach-type quartz sand mixed with small amounts of clays (primarily smectite, palygorskite, and sepiolite), black phosphate grains, and lime muds (calcilutites), and is generally a dark gray or gray-brown color. Subsequent research has shown that the Murdock Station Formation comprises two separate members: the lower **Jupiter Waterway Member** (early Zanclean) and the upper **Sarasota Member** (late Zanclean) (both informally proposed here as new members and described in the following sections).

The Murdock Station locally indurates into a dark gray phosphatic arenaceous limestone containing molds of bivalve and gastropod mollusks. At the Quality Aggregates quarries at Sarasota (Figure 3.9), the upper surface of the Sarasota Member (the top 0.5 m, in contact with the base of the overlying Buckingham Member of the Tamiami Formation), is indurated into a dark khaki-gray moldic arenaceous limestone. The Murdock Station differs from the Peace River Formation in having extremely thick, well-developed, and well-preserved beds of mollusks, barnacles, and ahermatypic corals, and these make up the majority of the carbonate fraction. Both members of the underlying Bayshore Formation differ in having thinner, less-developed shell beds (and a corresponding lower carbonate fraction) and containing a higher percentage of clay minerals.

As pointed out by Hunter (1968: 443), two distinct sets of beds are present in the Murdock Station: an upper set containing gray oyster–barnacle hash, fine to medium black phosphate grains, and clayey quartz sand, and a lower set, containing dense beds of large *Chesapecten* scallops, gray to brown clayey quartz sand with abundant phosphorite grains and phosphate-coated pebbles, and stringers of highly phosphatic unfossiliferous gray sand. Ketcher (1992), in her study of Unit 11 at Sarasota (see Figure 4.3 in Chapter 4), also recognized a lower scallop-rich zone and an upper barnacle-rich zone, but added an uppermost shell bed composed of *Diplodonta* clams. At Quality Aggregates pit #6 at Sarasota (Figure 3.9), the upper set of beds contained massive "reefs" (bioherms) of the giant barnacle *Concavus* and intertwined large bioherms of the branching aher-matypic coral *Septastrea crassa*. The lower set of beds ("Unit 12"; see Figure 4.3 in Chapter 4) contained dense aggregations of the giant scallops *Chesapecten jeffersonius* and *Chesapecten palmyrensis*, and the oyster *Ostrea compressirostra*.

FIGURE 3.9 View of the Quality Aggregates, Inc., pit #6, Fruitville area, Sarasota, Sarasota County, Florida, showing exposures down to depths of over 20 m. At the bottom of the quarry, near the pooled water in the foreground, the dark-colored phosphatic sediments of the Sarasota Member of the Murdock Station Formation are exposed (with massive barnacle bioherms). These can be seen to contact the light-colored sediments of the Buckingham Member of the Tamiami Formation at about 1.5 m above the water surface. The large, high pile of material above the Buckingham layer is composed of sediments from the Pinecrest Member of the Tamiami Formation (being removed for road fill). At the bottoms of the deeper drainage canals (seen in the distant upper left), the "Pecten Biostrome" of the Jupiter Waterway Member of the Murdock Station Formation was exposed.

The Murdock Station Formation appears to have a wider areal extent than does the underlying Bayshore Formation (Figure 3.10). Both members of the Murdock Station have the same distribution, and extend from Pinellas County in the north to Monroe and Dade Counties in the south, and throughout the Everglades area. At the Florida Rock Industries Naples Quarry (old "Mule Pen Quarry"), East Naples, Collier County, the Murdock Station was dredged from 20 m depth and contained large numbers of the oyster *Ostrea compressirostra* and small barnacles, all in a light gray phosphatic clay. The formation extends eastward into the Caloosahatchee and Kissimmee River valleys and has been identified in shallow (5–10 m depth) dredgings at the headwaters of the Kissimmee River, near old Fort Kissimmee, Highlands County (during Kissimmee wetland remediation excavations) and on spoil piles along the Kissimmee Prairie, Okeechobee County. Here, a gray phosphatic clay containing *Chesapecten palmyrensis* was exposed all along the canal dig and on spoil islands in the marshland. The phosphatic "Wabasso Beds" (Scott, 1988) encountered along coastal southeastern Florida (Figure 3.10), particularly in Indian River and St. Lucie Counties, appear to belong to the Murdock Station Formation and represent the extreme eastern formational development. Dated as Zanclean Pliocene (Scott, 1988: 84), the Wabasso Beds may be correlative with Hunter's "Upper Bed" of the Murdock Station (here referred to as the Sarasota Member).

Throughout its lateral extent, the Murdock Station is variable both in thickness and in depth below mean surface. At Sarasota, the upper formational surface (Unit 11, Figure 4.3) occurs at approximately 20 m below surface, and this same mean depth is seen at the "Mule Pen Quarry" at

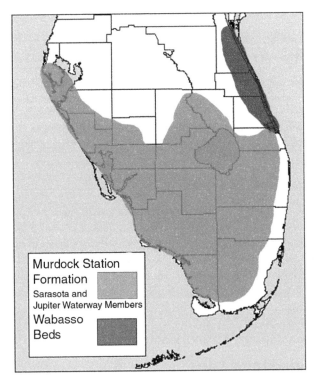

FIGURE 3.10 Areal distribution of the Murdock Station Formation and its two members, the Jupiter Waterway and the Sarasota. The "Wabasso Beds" facies can be seen to extend up the eastern coast.

East Naples. At the stratotype near Murdock, in the Port Charlotte area, the upper surface of the formation occurs at much shallower depths, from approximately 2.5 m to 3 m, and averages 2 m in thickness. The formation (both members) is over 8 m thick at Sarasota but may reach thicknesses of over 10 m at Naples. The Kissimmee Murdock Station beds are closer to mean surface and may be only 1 m thick.

STRATOTYPE

From Hunter (1968: 443): " … Canal bank, Sec. 4, T. 40 S., R. 21 E., on Jupiter Waterway, 1 mile south of U.S. Route 41, Port Charlotte, Charlotte County, Florida." The formation was described from the banks of the canal near the old railroad station at Murdock, where it was exposed 2 m above the water surface.

A classic stratigraphic column for the Murdock Station sequence was given by Hunter (1968: 443), and is reproduced as follows (with modifications by the authors in light of new data):

Bed	Description	Thickness (in feet)
	Pleistocene Series	
	Fort Thompson Formation	
8	Tan sand	1.0
7	Red-brown sand	2.0
6	Shelly sand containing *Pyrazisinus gravesae* (originally referred to as *Pyrazisinus scalatus*) and other species	3.0
5	Brown sand	1.0

Bed	Description	Thickness (in feet)

Pliocene Series (Originally Referred to as the Miocene Series)

Murdock Station Formation (originally placed in the Tamiami Formation as the Murdock Station Member)

4	Hard fragmental limestone with abundant barnacle fragments (upper bed of the Murdock Station)	3.0
3	Gray phosphatic sand	0.5
2	Pecten biostrome (containing *Chesapecten jeffersonius*) (lower bed of the Murdock Station)	0.5
1	Gray phosphatic sand	3.0
	Total depth	14

AGE AND CORRELATIVE UNITS

Based on the presence of key chronostratigraphic index fossils such as *Chesapecten jeffersonius* and *Ostrea compressirostra*, the Murdock Station Formation is now known to date from the Zanclean Pliocene. This uppermost formation of the Hawthorn Group is exactly correlative with the Sunken Meadow Member of the Yorktown Formation, Chesapeake Group ('Yorktown Zone 1") of Virginia and northern North Carolina, the lower beds of the Duplin Formation of southern North Carolina and northern South Carolina, and the lower beds of the Goose Creek Formation of southern South Carolina.

THE JUPITER WATERWAY MEMBER, MURDOCK STATION FORMATION

Lithologic Description and Areal Extent

This new member characteristically is composed of gray or brownish-gray clay (mostly sepiolite) with varying amounts of gray quartz sand and containing a high percentage of black phosphorite grains and phosphate-coated pebbles (3 to 5%; see Ketcher, 1992: 170). A high carbonate fraction is present in some facies, and this is almost always composed of scallop shell and oyster bioclasts. In some areas, such as near Port Charlotte, the classic clay facies interfingers with thick lenses of unfossiliferous phosphatic gray quartz sand ("Lower Bed"; Hunter, 1968: 443). In the APAC and Quality Aggregates quarries at Sarasota (Ketcher, 1992), the basal beds of the Jupiter Waterway Member are composed of heavily bioturbated unfossiliferous clays. The phosphatic "Alva Clay" and "LaBelle Clay" of Puri and Vernon (1964) are simply facies of the Jupiter Waterway Member. The upper bed ("Lower Zone" of Unit 11; see Ketcher, 1992) typically contains a thick, densely packed bed of large scallops, mostly *Chesapecten* species and the oyster *Ostrea compressirostra* (the "Pecten Biostrome" of Hunter, 1968).

The areal distribution of the Jupiter Waterway Member is the same as the entire formation, extending from Pinellas County south to Monroe and Dade Counties and underlying the entire Everglades. Throughout its range, the Jupiter Waterway can be easily traced by the presence of Hunter's "Pecten Biostrome" in canal digs, quarries, and driller's cores. Specimens of *Chesapecten jeffersonius*, *Chesapecten palmyrensis*, and *Ostrea compressirostra* were also found in deep well cores (70–80 m, in the area of the deep Loxahatchee Trough) from western Miami, Dade County (along U.S. 41), demonstrating that the Jupiter Waterway underlies the entire Everglades area. The new member varies in thickness from 1 m at the stratotype near Port Charlotte, Charlotte County, to 3 m at the "Mule Pen Quarry" at East Naples, Collier County, to over 10 m west of Miami, Dade County. A discernable unconformity is present at the top of the Jupiter Waterway Member, usually preserved as a thin sand or sandy limestone layer.

Stratotype

From Hunter (1968: 443): " … Canal bank, Sec. 4, T. 40 S., R. 21 E., on Jupiter Waterway, 1 mile south of U.S. Route 41, Port Charlotte, Charlotte County, Florida." The stratotype of the Jupiter

Waterway Member comprises Unit 2 ("Pecten Biostrome," type lower bed of the Murdock Station; Hunter, 1968) and Unit 1 ("gray phosphatic sand"; Hunter, 1968) of the stratigraphic column described in the previous section (the stratotype for the Murdock Station Formation).

Age and Correlative Units

Based on the presence of an entire assemblage of geochronologically restricted index fossils, the Jupiter Waterway Member is now known to date from the early Zanclean Pliocene. Some of these key early Zanclean markers include *Chesapecten jeffersonius*, *Euvola smithi*, *Carolinapecten yorkensis*, and *Ostrea compressirostra*. This phosphate-bearing unit is exactly correlative with the lower beds of the Sunken Meadow Member of the Yorktown Formation, Chesapeake Group (lower "Yorktown Zone 1") of Virginia and northern North Carolina, the basal beds of the Duplin Formation of southern North Carolina and northern South Carolina, and the basal beds of the Goose Creek Formation of southern South Carolina.

Jupiter Waterway Index Fossils

As the Jupiter Waterway sediments are heavily leached, only thick calcitic fossils are preserved. The following are some of the more indicative and commonly encountered species (Y = also found in Zone 1 of the Yorktown Formation; D = also found in the basal beds of the Duplin and Goose Creek Formations):

Gastropoda
Ecphora pachycostata (Y, D)
Ecphora roxaneae (Figure 3.11H) (D)
Bivalvia
Argopecten choctawhatcheensis (Figure 3.11G)
Carolinapecten yorkensis (Figure 3.11A) (Y)
Chesapecten jeffersonius (Figure 3.11D) (Y, D)
Chesapecten palmyrensis (Figure 3.11E) (D)
Euvola smithi (Figure 3.11F) (Y, D)
Leptopecten leonensis (Figure 3.11C)
Ostrea compressirostra (Figure 3.11B) (Y, D)

A single marine community from the Jupiter Waterway Member was described by the senior author (Petuch, 2004: 139–141), the *Chesapecten palmyrensis* Community (shallow sandy-bottom lagoon scallop beds).

THE SARASOTA MEMBER, MURDOCK STATION FORMATION

Lithologic Description and Areal Extent

From Petuch (1988: 120–121; originally described as the uppermost member of the "Hawthorn Formation"):

... Typically a yellow-tan calcareous clay, containing a large percentage of phosphorite pebbles and granules, and abundant phosphatized bone fragments. Bone fragments are mostly from large marine mammals such as cetaceans and sirenians. Shark teeth are also common. Exposed surfaces weather to a gray or gray-tan color. In some sections, the upper half meter is indurated into a hard, gray, phosphatic limestone, often highly fossiliferous.

Subsequent research at larger and better exposures (such as at Quality Aggregates pit #6 at Sarasota; Figure 3.9) has shown that the Sarasota Member is lithologically more diverse than

FIGURE 3.11 Index fossils for the Jupiter Waterway Member of the Murdock Station Formation. A = *Carolinapecten yorkensis* (Conrad, 1867), length 32 mm, juvenile specimen similar to holotype; B = *Ostrea compressirostra* Say, 1824, length 99 mm; C = *Leptopecten leonensis* (Mansfield, 1932), length 21 mm; D = *Chesapecten jeffersonius* (Say, 1824), length 142 mm; E = *Chesapecten palmyrensis* (Mansfield, 1936), length 83 mm; F = *Euvola smithi* (Olsson, 1914), length 72 mm; G = *Argopecten choctawhatcheensis* (Mansfield, 1932), length 36 mm; H = *Ecphora roxaneae* Petuch, 1991, length 55 mm. Differs from the younger Sarasota Member *Ecphora quadricostata* (Figure 3.12B) in having much wider and more rounded ribs that are ornamented with two to four incised grooves.

originally thought. The formal lithologic description is here expanded to include deposits of sandy clay (mostly smectite and sepiolite) with stringers of arenaceous calcilutites (sandy lime muds) and abundant black phosphate grains (3–5% phosphate; Ketcher, 1992). The largest part of the carbonate fraction is composed of bioherms of giant barnacles (Cirripedia) and densely intertwined beds of the branching ahermatypic coral *Septastrea* ("coral thickets" of Ketcher, 1992: 174). The calcilutite stringers often contain well-preserved specimens of aragonitic fossils, primarily small bivalves such as *Mulinia*, *Diplodonta*, and *Pleiorhytis*. Abundant marine mammal bones are typically present, usually from whales, walruses (Figure 3.13), seals, and sirenians (dugongs and manatees).

The Sarasota Member has essentially the same areal distribution as the underlying Jupiter Waterway Member. Over its range, the Sarasota can always be differentiated from the basal Murdock Station beds and from the Bayshore Formation by its higher percentage of carbonates (both lime muds and bioclasts) and by its lower percentage of clay minerals. The Sarasota Member reaches thicknesses of over 5 m at the "Mule Pen Quarry" at East Naples but only 2 m at the stratotype in the APAC pit at Sarasota (Figure 4.3, Chapter 4). The upper part of the "Wabasso Beds" of Indian River and St. Lucie Counties are probably correlative with the Sarasota. The new member is exposed at the stratotype of the Murdock Station Formation along the Jupiter Waterway. Here, it was designated Unit 4 ("Type Upper Bed") by Hunter (1968: 443) and was found to be only 1 m in thickness.

Stratotype

From Petuch (1988: 121):

> … The type is the basal bed (Unit 11 of Petuch, 1982) of the APAC pit mine (formerly the Macasphalt and Warren Brothers pit), Newburn Road, Sarasota, Sarasota County, Florida. In the APAC pit, only one meter or less of the Sarasota Member is exposed at 20 m depth. The new member is named for Sarasota County.

The APAC pit is located in Section 12, T. 36 S., R. 18 E., Sarasota County. The quarry is now flooded as a lake in a Sarasota County park and can be accessed by scuba diving.

Age and Correlative Units

Based on the presence of several key chronostratigraphic index fossils, such as *Ecphora quadricostata* and *Chesapecten jeffersonius* subspecies, the Sarasota Member is now known to date from the late Zanclean Pliocene. The new member is exactly correlative with the upper beds of the Sunken Meadow Member of the Yorktown Formation, Chesapeake Group of Virginia and northern North Carolina, the basal beds of the Duplin Formation of southern North Carolina and northern South Carolina, and the Goose Creek Formation of southern South Carolina. Of particular significance is the presence of a subspecies of *Chesapecten jeffersonius* (Figure 3.12E) that is the morphological intermediate between the older, classic Zanclean *C. jeffersonius* and the younger (early Piacenzian) *C. septenarius*. The morphology of this intermediate, as of yet unnamed subspecies, would indicate that the Sarasota Member dates from the very latest Zanclean, just prior to the Piacenzian. The indurated layer at the top of the Sarasota Member most likely represents a time of subaerial exposure (and erosion) at the Zanclean–Piacenzian boundary.

Sarasota Member Index Fossils

Although the Sarasota sediments are leached and the majority of fossils are calcitic, many aragonitic species have survived dissolution. These are often in a poorly preserved, fragile state, with the exception being the large branches of the ahermatypic coral *Septastrea crassa*. The following are typical Sarasota index fossils (Y = also found in Zone 1 of the Yorktown Formation):

FIGURE 3.12 Index fossils for the Sarasota Member of the Murdock Station Formation. A = *Concavus tamiamiensis* (Ross, 1965), clump length 152 mm; B = *Ecphora quadricostata* (Say, 1824), length 52 mm; differs from the older Jupiter Waterway Member *Ecphora roxaneae* (Figure 3.11H) in having thinner, more raised ribs that are "T"-shaped in cross-section and that are ornamented with a single deeply incised groove; C = *Discinisca multilineata* (Conrad, 1845), length 19 mm; D = *Pterorhytis umbrifer* (Conrad, 1832), length 78 mm. Note the large labial spine for opening barnacles; E = *Chesapecten jeffersonius* subspecies (with small *Concavus tamiamiensis* clump), length 101 mm. This unnamed form is intermediate between the older *Chesapecten jeffersonius* with 8 to 11 ribs and the younger *Chesapecten septenarius* with 5 to 7 ribs. The Sarasota subspecies differs from *C. septenarius* in having narrower, lower ribs with more rounded sides; F = *Placunanomia floridana* Mansfield, 1932, length 74 mm. Note the small hole that was probably drilled by an ecphora; G = *Planecphora hertweckorum* (Petuch, 1987), length 81 mm; H = *Balanus newburnensis* Weisbord, 1966, clump length 86 mm.

Gastropoda
Ecphora quadricostata (Figure 3.12B) (Y)
Planecphora hertweckorum (Figure 3.12G)
Pterorhytis umbrifer (Figure 3.12D) (Y)
Bivalvia
Chesapecten jeffersonius subspecies (intermediate between *C. jeffersonius* and *C. septenarius*)
  (Figure 3.12E)
Placunanomia floridana (Figure 3.12F)
Pleiorhytis centenaria (Y)
Margaritaria abrupta (Y)
Diplodonta subvexa (Y)
Mulinia congesta (Y)
Cirripedia
Balanus newburnensis (Figure 3.12H)
Concavus glyptopoma
Concavus sarasotaensis
Concavus tamiamiensis (Figure 3.12A) (originally thought to have come from the younger
  Tamiami Formation)
Brachiopoda
Discinisca multilineata (Figure 3.12C)
Cnidaria–Scleractinia
Septastrea crassa
Astrangia lineata

As can be seen in Figure 3.12E, valves of the unnamed subspecies of *Chesapecten jeffersonius* frequently acted as substrates for the settling of barnacle larvae. An examination in the field (APAC pit) of massive clumps of *Concavus tamiamiensis* revealed that many of the aggregations initially formed on the backs of *Chesapecten* valves. A single Sarasota Member marine community was described by the senior author (Petuch, 2004: 139–141): the *Concavus tamiamiensis* Community (intertidal mud flats and shallow lagoons).

Note: In the original description of the Sarasota Member (Petuch, 1988), the senior author included *Globecphora floridana* (as *Ecphora floridana*), *Ecphora whiteoakensis*, *Chesapecten middlesexensis*, and *Carolinapecten urbannaensis* (all species from the older Bayshore Formation) in the list of index fossils for the Zanclean-aged Sarasota Member. This mistake was due to the fact that specimens of these species were donated to the senior author by amateur fossil collectors, often with the erroneous locality data of "Unit 11 at the APAC quarry, Sarasota." Subsequent investigations have shown that these specimens (many illustrated here) actually were collected along the Cocoplum Canal and other areas near Port Charlotte and that they are restricted to the late Miocene.

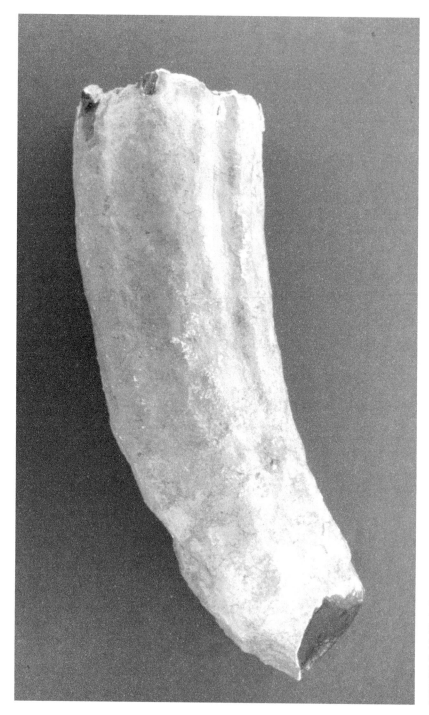

FIGURE 3.13 Tusk from an early Pliocene Florida Walrus (*Trichecodon*) from the Murdock Station Formation, Hawthorn Group. This 55-cm specimen was collected in Unit 11 in the APAC quarry at Sarasota in Sarasota County.

4 Middle and Late Pliocene Southern Florida

The middle and late Pliocene (early and late Piacenzian Age) saw the return of extended periods of warm, tropical climates and biogenic carbonate deposition on the Florida Platform. By the early Piacenzian, permanent upwelling systems had ceased off western Florida, and the Okeechobean Sea no longer received massive influxes of plankton blooms and nutrients. This oceanographic shift permanently altered the Okeechobean depositional environments. The corresponding change in water chemistry resulted in the cessation of phosphatic sedimentation and the initiation of pure carbonate and carbonate-quartz sand deposition. With the loss of upwelling-influenced water conditions in the early Piacenzian, the Upwelling-Deltaic Depositional Episode came to an end and the Pseudoatoll Depositional Episode was ushered in. This marks the beginning of the Okeechobee Group.

At the end of the Murdock Station deposition, a major regression caused the entire Okeechobean Sea to become emergent for over 200,000 years (Vail and Hardenbol, 1979; Haq et al., 1988; Dowsett and Cronin, 1990). During this time, subaerial exposure caused the upper surface of the Sarasota Member to indurate into a sandy phosphatic limestone. Following the latest Zanclean–earliest Piacenzian sea level drop, the Okeechobean Sea basin was again reflooded, but this time by deeper and warmer water (30 m above present sea level: Brooks, 1974; Dowsett and Cronin, 1990). This marine transgression, which may have resulted from an extended period of global warming, produced the Tamiami Subsea (named for the Tamiami Formation) (Figure 4.1).

During its span of over 2 million years, the Tamiami Subsea flooded four times, with the first two transgressions producing the deepest water conditions and lasting the longest (Figure 4.2). The last two transgressive intervals were much shorter in duration and were separated by longer periods of sea level lows. Judging from the gradual shallowing of each successive transgression, these last two Tamiami refloodings indicate the beginnings of global cooling and increased glaciation, presaging the Pleistocene. The eustatic fluctuations of the Tamiami Subsea are reflected in the four major depositional pulses seen in the Tamiami Formation, its principal stratigraphic unit. These led to the deposition of three sets of lithostratigraphic members, which are described in the following sections.

PALEOGEOGRAPHY OF THE TAMIAMI SUBSEA, OKEECHOBEAN SEA

The oyster–barnacle banks and deltas of Hawthorn Group time formed the base for the classic late Pliocene structural feature of the Okeechobean Sea, the **Everglades Pseudoatoll** (Figure 4.1; see Petuch, 1987, 1997, 2004). This immense coral structure and its associated carbonate environments grew upon, and expanded, the geomorphologic features formed during the late Miocene and early Pliocene. At its maximum development during late Piacenzian time, the Everglades Pseudoatoll was a "U"-shaped feature composed of zonated coral reef systems surrounding a broad, shallow carbonate platform covered with Turtle Grass beds, open sand areas, and patch reef coral bioherms. The reef systems extended from Charlotte County, around the periphery of the Tamiami Subsea, all the way northward to Palm Beach County, making this the largest known Pliocene coral reef structure. If stretched out straight, the "U"-shaped pseudoatoll would have extended over 330 km, making it longer than the Great Barrier Reef of the Recent Belize coastline. The Everglades Pseudoatoll, then, is the largest contiguous coral reef system known from the Cenozoic western Atlantic.

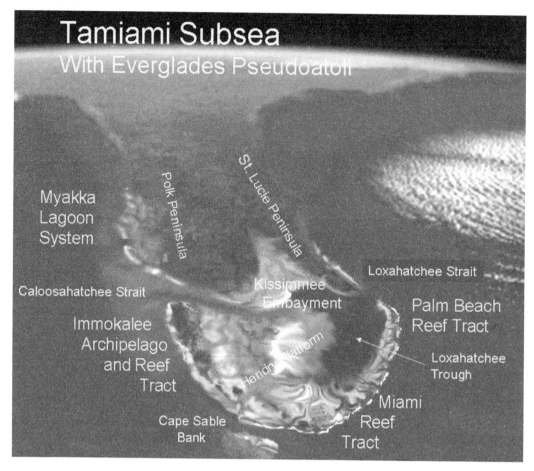

FIGURE 4.1 Simulated space shuttle image (altitude: 100 mi) of the Floridian Peninsula during the late Piacenzian Pliocene, showing the possible appearance of the Tamiami Subsea of the Okeechobean Sea and some of its principal geomorphological features.

Because of the structural complexity of the pseudoatoll and its proximity to the Floridian mainland, the Tamiami Subsea contained more types of tropical environments than were ever seen previously at any time in Florida's history. The main geographical feature, the pseudoatoll, contained the highest number of depositional environments and ecological zones. These were distributed in three main sections: the **Immokalee Reef Tract** (named for Immokalee, Collier County) in Lee, Collier, and Monroe Counties; the **Miami Reef Tract** (named for Miami, Dade County), in Monroe, Dade, and Broward Counties; and the **Palm Beach Reef Tract** (named for Palm Beach County). Each of these reef tracts had its own distinctive combination of coral zonational patterns and ecosystems. The Immokalee Reef Tract was the widest of the three sections and had developed a chain of large coral cays, the **Immokalee Archipelago**. These appear to have been covered with mangrove forests. A subsidiary carbonate system, the **Cape Sable Bank**, had now formed off the southwestern edge of the pseudoatoll.

Within the interior of the pseudoatoll, the dominant feature was the **Hendry Platform** (named for Hendry County). As the surrounding coral reef tracts had grown upon the Hawthorn oyster banks, the Hendry Platform covered the surface of the drowned Immokalee Delta. Sedimentation on the platform was complex, with the northern side receiving a large quartz sand component that derived from rivers draining off the peninsular highlands. The southern end of the Hendry Platform, being far from fluvial influences, received fewer siliciclastics and retained a higher percentage of

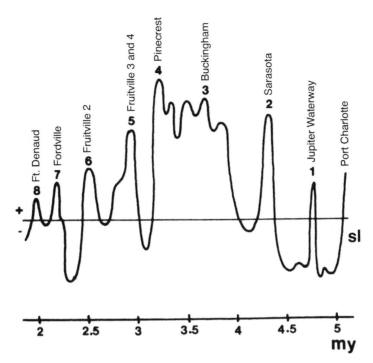

FIGURE 4.2 Sea level curve for the Pliocene, showing major eustatic fluctuations. Present sea level (sl) is used as a reference standard. Time is in millions of years (my). Numbers correspond to the times of deposition of formations and members, during sea level highs. For Formations: (1, 2) = Murdock Station Formation; (3, 4, 5, 6) = Tamiami Formation; (7, 8) = Caloosahatchee Formation. For Members: 1 = Jupiter Waterway Member, Murdock Station Formation; 2 = Sarasota Member, Murdock Station Formation; 3 = Buckingham Member, Tamiami Formation; 4 = Pinecrest, Ochopee, and Golden Gate (basal beds) Members, Tamiami Formation; 5 = Fruitville and Golden Gate Members (Unit 3 and Unit 4 equivalents), Tamiami Formation; 6 = Fruitville and Golden Gate Members (Unit 2 equivalent), Tamiami Formation; 7 = Fordville Member, Caloosahatchee Formation; 8 = Fort Denaud Member, Caloosahatchee Formation. (From the Oxygen Isotope Model of Krantz, D.E., 1991, *Quaternary Science Reviews,* 10: 163–174; Campbell, L.D., 1993, *Pliocene Molluscs from the Yorktown and Chowan River Formations of Virginia*, Virginia Division of Mineral Resources, Publication 127, Charlottesville, VA, 259 pp.)

carbonate sand and mud. The carbonate sedimentary input derived primarily from fines eroding off the Immokalee and Miami Reef Tracts. This lithofacies difference is readily seen in the contemporaneous Pinecrest and Ochopee Members of the Tamiami Formation (discussed later in this chapter). Here, the quartz-sand-rich Pinecrest was deposited in the northern and central areas of the bank whereas the carbonate-rich, low-sand Ochopee was deposited in the extreme south. To the east of the Hendry Platform, a deep trough feature, the **Loxahatchee Trough** (named for Loxahatchee, Palm Beach County), occupied the depression related to the Everglades Unconformity.

During Tamiami time, the southern end of Florida was split into two smaller peninsulas, which were separated by a wide, shallow embayment. The western peninsula, the **Polk Peninsula** (named for Polk County) was the widest and was fringed by extensive mangrove forests and estuarine environments. The largest of these was the **Myakka Lagoon System** (named for the Myakka River, Sarasota County), which housed the richest Pliocene tropical estuarine fauna known from anywhere in the western Atlantic. A similar system of estuarine environments existed within the large bay-like feature east of the Polk Peninsula, the **Kissimmee Embayment** (named for the Kissimmee River). The Kissimmee Embayment differed from the Myakka Lagoon System in having much larger and more extensive areas of open mud and sand flats. To the east of the Kissimmee Embayment, the smaller Floridian peninsula, the **St. Lucie Peninsula** (named for St. Lucie County), separated the Okeechobean Sea from the open Atlantic. Besides the innumerable small, shallow channels and cuts between the pseudoatoll reefs and islands, the Tamiami Subsea was connected to open oceanic conditions only through two narrow, deep channels; the **Caloosahatchee Strait** (named for the Caloosahatchee River) in the west, which connected to the Gulf of Mexico, and the **Loxahatchee Strait** (named for the Loxahatchee River) in the east, which connected to the Atlantic Ocean.

The exceptionally diverse environments of the Tamiami Subsea created habitats for an unusually rich molluscan fauna that was replete with endemic local adaptive radiations. Within most units of the Tamiami Formation, the remnants of these unique faunas exist as beautifully preserved, thick shell beds, often filling entire exposed sections. Many of the more interesting and stratigraphically important of these mollusks are illustrated in the following sections of this chapter, primarily in their role as index fossils for the members of the Tamiami Formation.

THE TAMIAMI FORMATION, OKEECHOBEE GROUP

HISTORY OF DISCOVERY

The Tamiami Formation, the principal stratigraphic unit of the Piacenzian Pliocene, has had a confused and ill-defined nomenclatural history. Because of the lack of surface exposures and because of the absence of collecting sites within the Everglades, the Tamiami Formation has had as many interpretations as there have been workers in southern Floridan geology. The name "Tamiami Limestone" was first proposed by Mansfield (1939, published posthumously) only for " … limestone penetrated in digging shallow ditches to form the road bed of the Tamiami Trail." This unit was later named the "Ochopee Limestone Member" by Muriel Hunter (1968) in her expanded nomenclatural scheme (discussed later in this section). In his 1939 paper, Mansfield also named the "Buckingham Limestone" from Lee County, considering it an older unit subjacent to his "Tamiami Limestone." Based on geochronologic guide fossils, Mansfield's original assignment of an older, subjacent Buckingham and a younger, suprajacent "Tamiami" (Ochopee) is now known to be correct. Parker et al. (1955) were the first workers to attempt an expanded definition of the "Tamiami" as a formation.

In the early 1930s, Tucker and Wilson described the "Acline Fauna" from small shell pits near Punta Gorda (1932, 1933). Much of the Acline molluscan fauna was later encountered by Axel Olsson in excavations during construction of "Alligator Alley" and along the Tamiami Trail (U.S. Route 41) in the early 1960s. He later (Olsson in Olsson and Petit, 1964: 516) named the sandy units housing this Acline-type fauna the "Pinecrest Beds," after the old settlement of Pinecrest on the Everglades Road, which branches off U.S. Route 41 at the 40-Mile Bend on the Dade–Collier County line. He later recognized this same sandy unit and accompanying index fossils at other places around the Everglades, such as along the Miami Canal in western Broward and Palm Beach Counties, at the Brighton Indian Reservation on the northwestern side of Lake Okeechobee ("Brighton Facies"), and in canal digs in Glades, Highlands, and Hendry Counties.

As more excavations occurred across the area in the 1970s and 1980s, geologists were given the opportunity to examine other facies of the Tamiami Formation. Some of these included coral reef tracts (the "Golden Gate Reef Member" of Missimer, 1992: 67; based on Meeder's unpublished dissertation) and lagoonal and estuarine facies (the "Fruitville Formation" of Waldrop and Wilson, 1990) from the Sarasota area. Earlier, Hunter (1968) expanded upon Parker's redefinition of the "Tamiami Formation," incorporating the Buckingham and Pinecrest with three new members, the previously mentioned Ochopee and two lower, phosphatic members, the Bayshore and Murdock Station. These last two "members" were removed from the Tamiami Formation and given full formational status by the senior author (Petuch, 2004; discussed in the previous chapter).

In 1982, the senior author published the first extensive measured section through a Tamiami ("Pinecrest Beds") exposure in the Ashland Petroleum and Asphalt Corp. (APAC) quarry in Sarasota.

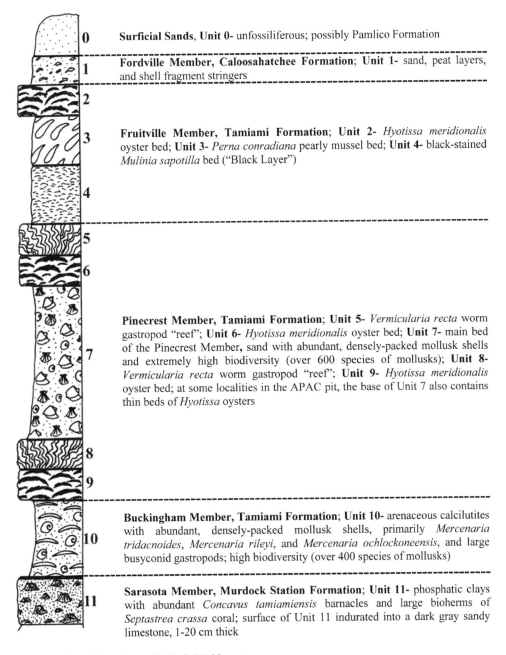

0 Surficial Sands, Unit 0- unfossiliferous; possibly Pamlico Formation

1 Fordville Member, Caloosahatchee Formation; Unit 1- sand, peat layers, and shell fragment stringers

2
3 Fruitville Member, Tamiami Formation; Unit 2- *Hyotissa meridionalis* oyster bed; Unit 3- *Perna conradiana* pearly mussel bed; Unit 4- black-stained *Mulinia sapotilla* bed ("Black Layer")
4

5 Pinecrest Member, Tamiami Formation; Unit 5- *Vermicularia recta* worm
6 gastropod "reef"; Unit 6- *Hyotissa meridionalis* oyster bed; Unit 7- main bed of the Pinecrest Member, sand with abundant, densely-packed mollusk shells and extremely high biodiversity (over 600 species of mollusks); Unit 8-
7 *Vermicularia recta* worm gastropod "reef"; Unit 9- *Hyotissa meridionalis* oyster bed; at some localities in the APAC pit, the base of Unit 7 also contains thin beds of *Hyotissa* oysters

8
9

10 Buckingham Member, Tamiami Formation; Unit 10- arenaceous calcilutites with abundant, densely-packed mollusk shells, primarily *Mercenaria tridacnoides*, *Mercenaria rileyi*, and *Mercenaria ochlockoneensis*, and large busyconid gastropods; high biodiversity (over 400 species of mollusks)

11 Sarasota Member, Murdock Station Formation; Unit 11- phosphatic clays with abundant *Concavus tamiamiensis* barnacles and large bioherms of *Septastrea crassa* coral; surface of Unit 11 indurated into a dark gray sandy limestone, 1-20 cm thick

Entire stratigraphic column (Units 0-11) 20 meters

FIGURE 4.3 Stratigraphic column from the eastern side of the APAC pit (formerly Richardson Quarry and Warren Brothers pit), Fruitville area of Sarasota, Sarasota County, Florida. Three formations were exposed within the quarry; the Caloosahatchee Formation (Fordville Member, Unit 1), the Tamiami Formation (Fruitville, Pinecrest, and Buckingham Members, Unit 2 to Unit 10), and the Murdock Station Formation (Sarasota Member, Unit 11). A Unit 12 (Murdock Station Formation, Jupiter Waterway Member) was later exposed at the bottoms of deep drainage channels (gray phosphatic clays with dense beds of large scallops). Unit 11 at this quarry is the stratotype of the Sarasota Member of the Murdock Station Formation. (Modified from Petuch, E.J., 1982, *Proceedings of the Academy of Natural Sciences of Philadelphia*, 134: 12–30.)

In that paper, 11 stratigraphic units were defined and numbered, with paleoenvironmental and biostratigraphic descriptions being given for each unit. Originally, three main units were recognized: an upper Caloosahatchee Formation (Unit 1), a middle "Pinecrest Formation" (Units 2–10), and a lower "Hawthorn Formation" (Unit 11). Since that time, with a greatly increased quality of geo-chronological, biostratigraphic, and paleoenvironmental data from around the Everglades area, we have been able to improve the original Sarasota stratigraphic column (shown here in Figure 4.3). As now understood, only Units 5–9 represent the Pinecrest, with Unit 11 representing the top of the Hawthorn Group (Sarasota Member of the Murdock Station Formation), Unit 10 representing the Buckingham, Units 2–4 representing the Fruitville, and Unit 1 representing the Fordville Member of the Caloosahatchee Formation (see Chapter 5). This section can be used as a standard reference section for the Pliocene of Florida and the Atlantic Coastal Plain.

Of all the units and faunizones described by previous workers, we recognize only five nomen-clatural subdivisions for the Tamiami Formation: the lower Buckingham Member, the middle Pinecrest and Ochopee Members, and the upper Fruitville and Golden Gate Members. Our definition is a compromise (updated with new and more accurate data) between the schemes of Hunter (1968), Waldrop and Wilson (1990), Lyons (1991), and Missimer (1992).

LITHOLOGIC DESCRIPTION AND AREAL EXTENT

Following Hunter's and Missimer's descriptions, with modifications, we here define the Tamiami Formation as consisting of variably interbedded calcarenites, calcilutites, quartz sands, and limestones and being devoid of massive phosphate deposition (except for reworked Murdock Station grains found in some facies of the Buckingham Member). Typically, many of the unconsolidated beds contain dense accumulations of bioclasts, mostly large mollusks and corals. Many of these shell beds, especially those composed mainly of bivalves, are so closely packed and dense that there is only a tiny fraction of any type of sediment matrix present between bioclasts. Clays are rarely present, being confined primarily to shoreline and estuarine facies, and always in minor quantities. The calcilutites are typically white or light gray, whereas the calcarenites and limestones are generally light gray, often with variable amounts of quartz sand intermixed. Quartz sand beds vary in color from white to dark gray, with some being stained yellow, brown, or light yellow-orange. The coral and mollusk bioclasts, when in fresh condition, are generally extremely well preserved and are usually stained a brown, tan, or bluish-gray color. Mollusk specimens from reducing environments in estuarine facies are stained black or dark blue-gray. Both aragonitic and calcitic fossils are typically present, although some limestone units are leached and contain only calcitic specimens.

The Tamiami Formation underlies the entire Everglades region (best seen in the distributional map of the Pinecrest and Ochopee Members on Figure 4.8), extending from Pinellas County in the northwest to Monroe County in the southwest and from southernmost Osceola County (Okeechobee Plain) south to Dade County. Throughout its entire range, there are no natural surface outcrops of the Tamiami. The formation ranges in thickness from 15 m at Sarasota, Sarasota County (Figure 4.3), to over 37 m at East Naples, Collier County (Missimer, 1992: 87–89), to over 60 m under the Atlantic Coastal Ridge (Swayze and Miller, 1984), to only 7 m in the Kissimmee Valley near Fort Drum, Okeechobee County. The surface contours of the Tamiami Formation follow the bathymetric contours of the Tamiami Subsea.

STRATOTYPE

Although the name "Tamiami" has been used for this completely redefined and expanded formation, it is still necessary to give Mansfield's original type locality (from Mansfield, 1939: 8): " ... a limestone penetrated in digging shallow ditches to form the road bed of the Tamiami Trail over a distance of 34 mi in Collier and Monroe Counties, Florida." Because this section is really repre-sentative of only one (the Ochopee) of the five members of the Tamiami Formation, we suggest that the measured section from the APAC pit in Sarasota (Figure 4.3) be added to Mansfield's stratotype as a standard reference section.

AGE AND CORRELATIVE UNITS

Based upon entire suites of geochronologic index fossils, the Tamiami Formation is now known to span most of the late Pliocene, from the earliest Piacenzian to the latest Piacenzian. The presence of diagnostic species of widespread molluscan groups, such as the bivalves *Carolinapecten*, *Nodipecten*, and *Mercenaria* and the gastropods *Terebraspira*, *Latecphora*, and *Contraconus* (discussed in the following sections), show that the formation is exactly correlative with the Rushmere, Mogarts Beach, and Moore House Member of the Yorktown Formation of Virginia and northern North Carolina, and the Edenhouse Member of the Chowan River Formation of the same area. The Tamiami is also correlative with the middle and upper beds of the Duplin Formation of southern North Carolina and northern South Carolina, the middle and upper beds of the Goose Creek Formation of South Carolina, and the Jackson Bluff Formation of the Florida Panhandle.

TAMIAMI INDEX FOSSILS

The single most important index fossil group within the Tamiami Formation is the gastropod family Cypraeidae, the cowrie shells. During the Piacenzian Pliocene, cowries underwent amazing species radiations within the Okeechobean Sea, producing whole suites of endemic, chronologically restricted species. There are 37 described species (with at least 20 other as-of-yet undescribed species) encompassing 4 genera, 2 described subgenera, and several undescribed subgenera that evolved within the Tamiami Subsea, making this the single largest radiation of cowries ever found in one place anywhere on Earth (Petuch, 1994, 1996, 1998, 2004). With the exception of two widespread eastern American species (*Akleistostoma carolinensis* and *A. pilsbryi*), all the members of these species radiations were confined to the area of the Okeechobean Sea. The Florida cowries were gregarious, occurring in huge aggregations numbering in the tens of thousands (much like the living mouse cowrie *Muracypraea mus* of Venezuela, a distant relative with similar ecological preferences (see Petuch, 1976; 1979; 1988). Their abundance, large size, and rapid evolution make them ideal stratigraphic index fossils. All 37 described Tamiami species are illustrated in the following sections and are arranged by stratigraphic position. Other important Tamiami index fossil groups, which can be used to demarcate the boundaries between members, include the gastropod genera *Hystrivasum*, *Terebraspira*, *Echinofulgur*, and *Pterorhytis* and the bivalve genera *Carolinapecten* and *Nodipecten*.

THE BUCKINGHAM MEMBER, TAMIAMI FORMATION

History of Discovery

Although originally recognized by Matson and Clapp (1909) and Cooke and Mossom (1929), the outcrops of limestone exposed in the Caloosahatchee River channelizing excavations and in quarries near Buckingham, Lee County, were assigned to the "Choctawhatchee Marl." It was not until 1939 that Mansfield formally named the unit (as the "Buckingham Limestone") and gave a detailed list of molluscan index fossils. While the paper was in press (published September 1, 1939), Wendell Mansfield died (July 24) having seen the galley proofs before his death but not the final page proofs. Later, Cooke (1945: 210) changed the name to "Buckingham Marl," basing this refinement on work that he had done a year earlier with Gerald Parker. They noticed that the exposed, weathered surface indurates into a limestone but that the unexposed material is actually an unconsolidated fossiliferous calcarenite.

Hunter (1968) placed the Buckingham within her expanded concept of the Tamiami Formation, in this case as the "Buckingham Limestone Member." Based on the misidentification of some scallop species (*Nodipecten* and the *Christinapecten tamiamiensis* species complex), Hunter considered the Buckingham to be time-equivalent with the Ochopee and Pinecrest Members. Missimer (1992: Figure 3.3) correctly identified the Buckingham as being subjacent to the Ochopee Member and as being the oldest and lowest part of the Tamiami sequence.

Lithologic Description and Areal Extent

The Buckingham Member was originally only sketchily described (from Mansfield, 1939: 8, 12):

> A new formation is here proposed for a limestone cropping out in Lee County, Florida … The molluscan fauna of the Buckingham limestone consists mainly of *Pecten* and *Ostrea*, which are well-preserved; but most of the other genera are preserved only as casts and molds … The matrix in which the fossils are embedded consists of a chalky limestone that contains little sand and many small grains of brown phosphorite. The rock hardens on exposure and changes to a brownish color.

Cooke (1945: 210) further expanded the lithostratigraphic definition:

> Fresh exposures of the Buckingham marl consist of cream-colored impermeable calcareous clay containing small grains of a brown phosphatic mineral, which increase in abundance toward the lower part of the formation. The Buckingham is fairly uniform throughout its known extent, but where it has long been exposed to weathering, the marl hardens into firm limestone.

Hunter's description essentially follows Cooke's:…

> Light gray to white soft calcareous clay (a calcilutite) that weathers to a buff color. It contains some quartz sand, a few bone fragments and shark teeth, and few grains of brown phosphate. Poorly preserved fossil molds and well-preserved pectens, oysters, barnacles, echinoids, etc., are present.

The most interesting aspect that is mentioned in all three descriptions is the presence of a "brown phosphorite." As Missimer (1992: 68) pointed out, these phosphatic mineral grains are reworked from the underlying Hawthorn Group. These would have eroded off the indurated surface of the Murdock Station Formation when the Everglades area was reflooded by the Tamiami Subsea. These brown (and etched) phosphate grains were later mixed in with the younger Buckingham sediments. The gradational content of the phosphate component pointed out by Cooke further indicates that these grains were reworked from older beds and were not part of Buckingham deposition.

The best and largest exposures of the Buckingham Member were in the Quality Aggregates pit #6 (Figure 3.9, Chapter 3) and at the APAC pit (Figure 4.3), both at Sarasota. At these two localities, the Buckingham was composed of a sandy lime mud (arenaceous calcilutite) with abundant, closely packed mollusk shell bioclasts, mostly large bivalves such as *Mercenaria tridacnoides* and *Carolinapecten walkerensis*. As at the stratotype, exposed surfaces often weathered into a soft arenaceous limestone, closely resembling that at Buckingham. Blocks of a nonphosphatic arenaceous limestone, containing perfect specimens of *Mercenaria tridacnoides*, were also brought up by draglines at Mule Pen Quarry, East Naples, Collier County, and at the Bird Road lake excavation in Miami, Dade County.

The Buckingham Member has the same areal distribution as does the entire Tamiami Formation, with the exception of Pinellas County, where the member appears to be absent (Figure 4.4). Throughout its distribution, the member is variable in thickness, ranging from 7 m thick near Alva, Lee County, to 2.5 m thick at Sarasota, Sarasota County, and to over 8 m thick west of Miami, Dade County. The surface contours of the Buckingham Member follow the bathymetric contours of the Tamiami Subsea.

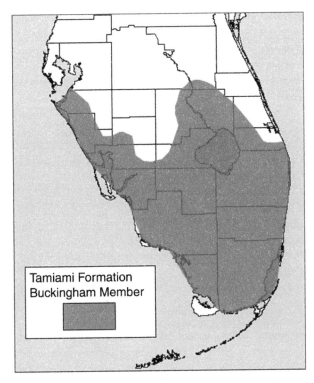

FIGURE 4.4 Areal distribution of the Buckingham Member of the Tamiami Formation.

Stratotype

From Mansfield (1939: 8): " ... The type locality is at a quarry near State Highway 25, half a mile west of Orange River, Lee County, Florida (Sec. 5, T. 44 S., R. 26 E.)." The APAC section (Figure 4.3) can be used as a standard reference section for the Buckingham, showing the suprajacent and subjacent units.

Age and Correlative Units

Based on the presence of key stratigraphic index fossils, such as *Akleistostoma carolinensis*, *Planecphora mansfieldi*, *Mercenaria tridacnoides*, *Nodipecten peedeensis*, and *Lirophora ulocyma*, the Buckingham Member is now known to date from the earliest Piacenzian Pliocene. The member is exactly correlative with the Rushmere Member of the Yorktown Formation of Virginia and northern North Carolina, the lower beds of the Duplin Formation of southern North Carolina and northern South Carolina, the lower beds of the Goose Creek Formation of South Carolina, and the lower member (unnamed, the *Ecphora* Zone at Alum Bluff) of the Jackson Bluff Formation of the Florida Panhandle (Mansfield, 1930).

Buckingham Index Fossils

The Buckingham molluscan fauna is one of the richest known from the Okeechobee Group. The highly distinctive and characteristic assemblages found in this member contain three faunal components: an endemic Okeechobean Sea component, a component of species shared with the "*Ecphora* Zone" of the Jackson Bluff Formation (the Jackson Subsea of the Choctaw Sea; see Petuch, 2004), and a component of widespread species shared with the Duplin and Goose Creek Formations. Some of the more abundant and indicative include:

Endemic Buckingham species (Tamiami Subsea restricted)
Gastropoda
Apicula buckinghamensis (Figure 4.7E)
Akleistostoma crocodila (Figure 4.6K)
Siphocypraea (Pahayokea) erici (Figure 4.6I) (oldest known species of *Pahayokea*)
Calusacypraea duerri (Figure 4.6F)
Sconsia metae
Busycon pachyus
Lindafulgur lindajoyceae (Figure 4.6A)
Pyruella waltfrancei (Figure 4.7B)
Sinistrofulgur grabaui (Figure 4.7F)
Echinofulgur cannoni (Figure 4.7F)
Tropochasca petiti (Figure 4.5J)
Fusinus dianeae (Figure 4.6D)
Hystrivasum barkleyae (Figure 4.7I)
Hystrivasum violetae
Oliva (Strephona) keatoni
Scaphella martinshugari
Seminoleconus violetae (Figure 4.7C)

Species shared with the "*Ecphora* Zone" of the Jackson Bluff Formation (see Mansfield, 1930, 1932)
Gastropoda
Diodora (Glyphis) alumensis
Torcula alumensis (Figure 4.7G)
Ecphora quadricostata striatula (Figure 4.5I)
Phyllonotus leonensis
Pterorhytis marshalli (Figure 4.5H)
Heilprinia dalli
Heilprinia gunteri
Busycoarctum tudiculatum (Figure 4.6C)
Busycotypus libertiense (Figure 4.5B)
Brachysycon propeincile
Pyruella harasewychi (Figure 4.6E)
Sinistrofulgur hollisteri Figure 4.5K)
Oliva (Strephona) alumensis
Oliva (Strephona) duerri
Ventrilia alumensis
Jaspidiconus harveyensis
Contraconus lindajoyceae (Figure 4.5F)
Bivalvia
Argopecten jacksonensis
Carolinapecten walkerensis (Figure 4.5C)
Euvola ochlockoneensis
Mercenaria ochlockoneensis (Figure 4.7J)
Lirophora ulocyma (Figure 4.7K)

Species shared with the Duplin and Goose Creek Formations
Eichwaldiella holmesi (Figure 4.6G)
Akleistostoma carolinensis (Figure 4.5G)
Akleistostoma pilsbryi (Figure 4.7H)

FIGURE 4.5 Index fossils for the Buckingham Member, Tamiami Formation. A = *Latecphora violetae* (Petuch, 1988), length 98 mm; B = *Busycotypus libertiense* (Mansfield, 1930), length 285 mm; C = *Carolinapecten walkerensis* (Tucker, 1934), length 96 mm (*Carolinapecten buckinghamensis* [Mansfield, 1939] is a synonym); D = *Mercenaria tridacnoides* (Lamarck, 1818), length 122 mm; E = *Planecphora mansfieldi* (Petuch, 1989), length 85 mm; F = *Contraconus lindajoyceae* Petuch, 1991, length 87 mm; G = *Akleistostoma carolinensis* (Conrad, 1841), length 60 mm; H = *Pterorhytis marshalli* (Mansfield, 1930), length 42 mm; I = *Ecphora quadricostata striatula* Petuch, 1986, length 57 mm; J = *Tropochasca petiti* Olsson, 1967, length 85 mm; K = *Sinistrofulgur hollisteri* Petuch, 1994, length 187 mm.

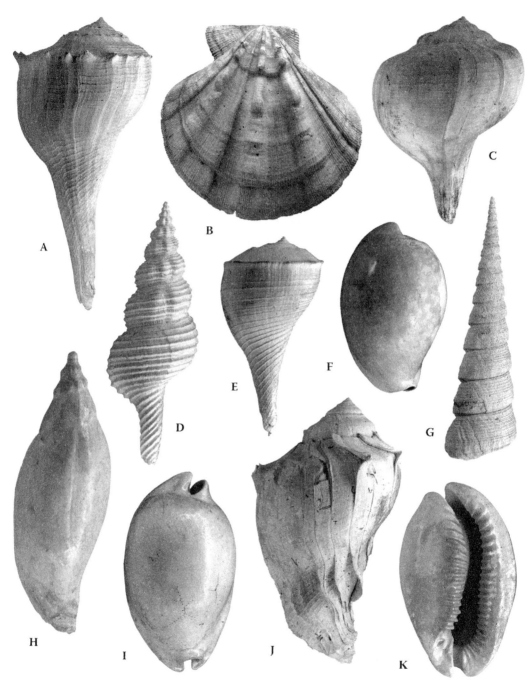

FIGURE 4.6 Index fossils for the Buckingham Formation, Tamiami Formation. A = *Lindafulgur lindajoyceae* (Petuch, 1991), length 131 mm; B = *Nodipecten peedeensis* (Tuomey and Holmes, 1856), length 118 mm; C = *Busycoarctum tudiculatum* (Dall, 1890), length 94 mm; D = *Fusinus dianeae* Petuch, 1994, length 108 mm; E = *Pyruella harasewychi* Petuch, 1982, length 73 mm; F = *Caluscypraea duerri* (Petuch, 1996), length 67 mm; G = *Eichwaldiella holmesi* (Dall, 1900), length 109 mm; H = *Volutifusus emmonsi* Petuch, 1994, length 110 mm; I = *Siphocypraea (Pahayokea) erici* Petuch, 1998, length 74 mm (oldest-known *Pahayokea*; listed as an *Akleistostoma* species in Petuch [2004: plate 54]); J = *Busycon filosum* (Conrad, 1862), length 133 mm; K = *Akleistostoma crocodila* (Petuch, 1994), length 64 mm.

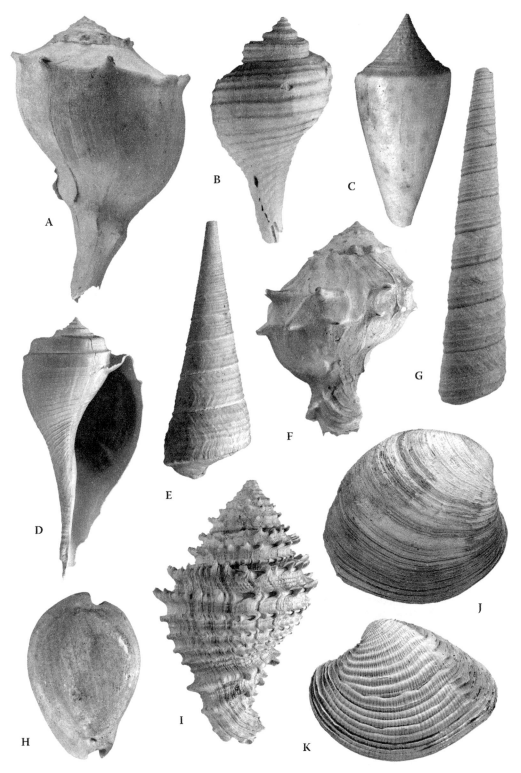

FIGURE 4.7 Index fossils for the Buckingham Member, Tamiami Formation. A = *Sinistrofulgur grabaui* Petuch, 1994, length 193 mm; B = *Pyruella waltfrancei* Petuch, 1994, length 56 mm; C = *Seminoleconus*

continued

FIGURE 4.7 (continued) *violetae* (Petuch, 1991), length 53 mm (note bands of spots and coronated spire); D = *Fulguropsis carolinensis* (Tuomey and Holmes, 1856), length 121 mm; E = *Apicula buckinghamensis* (Mansfield, 1939), length 65 mm (holotype is a broken, deformed cast and not shaped like typical specimens); F = *Echinofulgur cannoni* Petuch, 1994, length 69 mm; G = *Torcula alumensis* (Mansfield, 1930), length 143 mm; H = *Akleistostoma pilsbryi* (Ingram, 1947), length 26 mm; I = *Hystrivasum barkleyae* Petuch, 1994, length 88 mm; J = *Mercenaria ochlockoneensis* (Mansfield, 1932), length 86 mm; K = *Lirophora ulocyma* (Dall, 1895), length 46 mm.

Ficus holmesi
Latecphora violetae (Figure 4.5A)
Planecphora mansfieldi (Figure 4.5E)
Fulguropsis carolinensis (Figure 4.7D)
Busycon filosum (Figure 4.6J)
Volutifusus emmonsi (Figure 4.6H)
Bullata antiqua
Ventrilia smithfieldensis
Myurellina unilineata
Bivalvia
Nodipecten peedeensis (Figure 4.6B)
Mercenaria rileyi
Mercenaria tridacnoides (Figure 4.5D)
Marvacrassatella meridionalis

The marine communities of the Buckingham Member were described by the senior author (Petuch, 2004: 166–171) and include the *Mercenaria tridacnoides* Community (shallow muddy sand areas and intertidal mud flats), the *Apicula buckinghamensis* Community (shallow lagoon turritella beds), the *Nodipecten peedeensis* Community (deep lagoon pectinid beds), and the *Phyllangia blakei* Community (lagoonal coral bioherms). Several ahermatypic corals are also shared with the Jackson Bluff Formation (illustrated in Petuch, 2004: Plate 54) and include *Astrangia leonensis*, *Phyllangia blakei*, and *Paracyathus vaughani*.

THE PINECREST MEMBER, TAMIAMI FORMATION

History of Discovery

When Olsson and Petit (1964: 516–518) formally proposed the name "Pinecrest Beds," he really defined a molluscan biozone and not a lithostratigraphic unit. This was later restricted by Hunter (1968: 443) in her expanded definition of the Tamiami Formation. In this work, in which Hunter named the unit the "Pinecrest Sand Member," the first lithologic description was given for Olsson's widespread shell beds and their characteristic fauna. A more detailed description of the Pinecrest, showing a wider range of lithologies and paleoenvironments, was given by the senior author (Petuch, 1982) in a study of the APAC quarry at Sarasota (see Figure 4.3). Later, Missimer (1992: 65) gave more details regarding the areal distribution of what he called the "Pinecrest Sand Member." In this book, we refer to this middle Tamiami unit as simply the Pinecrest Member.

To paleontologists, the molluscan fauna of the Pinecrest Member is particularly noteworthy in that it is the richest known in the Okeechobee Group and the Pliocene of the Atlantic Coastal Plain. These mollusks lived during the warmest and most tropical time of the entire Pliocene (Figure 4.2), and over 600 species have been documented from this single unit alone. This number may rise to over 1000 when the micromolluscan fauna is better studied. Some of the most unusual and exotic fossil shells found anywhere in Florida have been described from the Pinecrest Member, and examples of these are shown in Figure 4.10 to Figure 4.14.

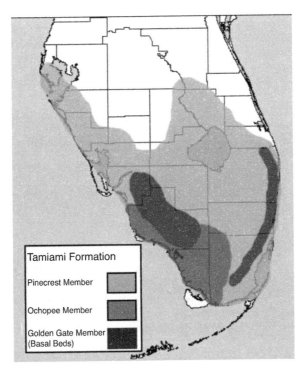

FIGURE 4.8 Areal distribution of the Pinecrest, Ochopee, and Golden Gate (basal beds) Members of the Tamiami Formation.

Lithologic Description and Areal Extent

As pointed out by Hunter (1968), the Pinecrest Member is composed of well-sorted, clean quartz sand, locally quite shelly, with abundant well-preserved marine fossils. From the stratigraphic column and descriptions shown in Figure 4.3, some of the Pinecrest shell beds are so dense and tightly packed that only a small amount of quartz sand exists between the large molluscan bioclasts. Some of these shell beds are essentially monocultures, being made up almost entirely of either *Hyotissa meridionalis* oysters, *Carolinapecten eboreus* scallop beds (Figure 4.9), *Vermicularia recta* worm gastropods (Figure 4.13), or the small strombid gastropod *Strombus floridanus*. The quartz sand component is usually white, pale yellow, or light gray in color. Variable amounts of lime muds (calcilutites) are often mixed with the quartz sand. The calcilutite/quartz sand ratio is gradational, with a higher quartz sand content in areas adjacent to the paleopeninsular mainland and a lower quartz sand content in areas adjacent to the pseudoatoll reefs and the Loxahatchee Trough. The shell bioclasts, when fresh, are stained a cream, tan, or light gray color.

The areal extent of the Pinecrest Member (Figure 4.8) is equivalent to that of the entire Tamiami Formation, except for southern Dade and northern Monroe Counties, where it interfingers with the contemporaneous Ochopee Member, and under the Immokalee Rise and the Atlantic Coastal Ridge, where it interfingers with the reefal deposits of the lower part of the Golden Gate Member. Thicknesses of the Pinecrest Member are variable, ranging from 10 m at Sarasota, to 2 m in southern Lee County, to over 7 m at Fort Drum, Okeechobee County, to over 20 m west of the Atlantic Coastal Ridge in Palm Beach County. The surface contours of the Pinecrest Member conform to the bathymetric contours of the Tamiami Subsea.

Within the Okeechobee Plain and Kissimmee River Valley, the lithology of equivalent beds differs from other areas of Pinecrest deposition. Here, high percentages of mud, clay minerals, and particulate organic matter are mixed with the quartz sand and shell bioclast components. This separate, distinct lithology, here referred to as the **Kissimmee Facies** of the Pinecrest Member,

FIGURE 4.9 View of the main shell bed (Unit 7) of the Pinecrest Member of the Tamiami Formation in the Quality Aggregates pit #6, Sarasota, Sarasota County, Florida. Here, a bed of the giant scallop *Carolinapecten eboreus* can be seen to be intercalated between layers of densely packed molluscan bioclasts, mostly the oyster *Hyotissa meridionalis* and the gastropod *Strombus floridanus*. This type of shell bed is typical of the Pinecrest Member. Note the waterfall on the far right (groundwater pouring into the pumped quarry). (Photo by Brian Schnirel.)

formed within the extensive mangrove forest and mud flat environments of the mid-Piacenzian Kissimmee Embayment. The unique environments of this area housed a highly endemic molluscan fauna, many of which are still unnamed and unstudied. Examples of some of these Kissimmee endemics are shown in Figure 4.14.

Stratotype

Olsson (1964: 516) notes that the Pinecrest Member is named

> … for certain strata composed largely of sand, barren or highly fossiliferous, encountered directly below a surface limestone in the general region of the 40-Mile Bend on the Tamiami Trail (Route 41) west of Miami in the western part of Dade County and extending across its boundary into Collier County, Florida. The name is taken from an old settlement on the Everglades Road (which branches off the present highway at 40-Mile Bend) about 1 mi west of the Dade–Collier County line.

The APAC section from Sarasota (Figure 4.3) can be used as a standard reference section for the Pinecrest Member, showing the suprajacent and subjacent beds.

Age and Correlative Units

Based on the presence of key widespread stratigraphic molluscan index fossils, such as *Latecphora bradleyae*, *Carolinapecten eboreus*, *Argopecten comparilis*, *Contraconus adversarius*, and *Terebraspira sparrowi*, the Pinecrest Member is now known to date from the mid-Piacenzian Pliocene.

The member is exactly correlative with the Mogarts Beach Member of the Yorktown Formation, Chesapeake Group of Virginia and northern North Carolina, the main beds of the Duplin Formation of southern North Carolina and northern South Carolina, the upper beds of the Goose Creek Formation of South Carolina, and the upper (unnamed) member (the "*Cancellaria* Zone") of the Jackson Bluff Formation of the Florida Panhandle (Mansfield, 1930, 1932).

Pinecrest Index Fossils

Unlike the fauna of the Buckingham Member, which contains many lower Jackson Bluff taxa, the Pinecrest fauna contains a much higher percentage of endemic taxa, approaching 60%. The other components comprise species shared with the Yorktown, Duplin, and upper beds of the Jackson Bluff Formation. Some of the more abundant and indicative include:

Endemic Pinecrest species (Tamiami Subsea restricted)
Gastropoda
Tegula lindae
Turbo (Marmarostoma) lindae (Figure 4.12D)
Apicula gladeensis (Figure 4.12H)
Eichwaldiella magnasulcus (Figure 4.11H)
Eichwaldiella pontoni (Figure 4.11K)
Vermicularia recta (Figure 4.11G, Figure 4.13)
Akleistostoma floridana (Figure 4.10F)
Siphocypraea (Pahayokea) mansfieldi (Figure 4.11B)
Calusacypraea globulina (Figure 4.12A)
Calusacypraea (Myakkacypraea) briani (Figure 4.11J)
Pseudadusta hertweckorum (Figure 4.11F)
Siphocypraea trippeana (Figure 4.10C)
Macrostrombus hertweckorum (Figure 4.12L)
Acantholabia sarasotaensis
Chicoreus stephensae (Figure 4.12G)
Chicoreus xestos (Figure 4.12B)
Hexaplex hertweckorum (Figure 4.10K)
Pterorhytis fluviana (Figure 4.10H)
Echinofulgur dalli
Tropochasca lindae (Figure 4.11E)
Melongena taurus (Figure 4.10B)
Cymatophos lindae (Figure 4.10G)
Busycon titan
Busycoarctum tropicalis
Busycotypus bicoronatum
Pyruella demistriatum
Pyruella rugosicostata (Figure 4.10D)
Parametaria hertweckorum
Turbinella streami
Hystrivasum locklini (Plate 4.10E)
Pleioptygma lindae
Trigonostoma druidi
Ventrilia senarium
Dauciconus bassi (Figure 4.12K)
Eugeniconus paranobilis (Figure 4.12E)
Jaspidiconus marymansfieldae

Lithoconus druidi (Figure 4.11A)
Spuriconus cherokus (Figure 4.11D)
Cymatosyrinx aclinica
Bivalvia
Caloosarca notoflorida
Nodipecten collierensis (*N. floridensis*, Tucker and Wilson, 1932, is a synonym) (Figure 4.11L)
   (also found in the contemporaneous Ochopee Member; see Figure 4.15A)
Venericardia olga

Species shared with the "*Cancellaria* Zone" of the Jackson Bluff Formation
Strombus floridanus (Figure 4.11I)
Dallitesta coensis
Cassis floridensis
Cymatium (Linatella) floridanum
Ficus jacksonensis
Seminoleconus trippae
Bivalvia
Chione procancellata (Figure 4.12J)

Species shared with the Duplin (D) and Yorktown (Y) Formations
Tegula exoluta (D)
Sconsia hodgei (D,Y)
Chicoreus floridanus (D) (Figure 4.12I)
Phyllonotus globosus (D) (Figure 4.11C)
Trossulasalpinx trossulus (D,Y)
Vokesinotus lepidotus (D,Y)
Ecphora quadricostata rachelae (D, Y)
Latecphora bradleyae (D) (Figure 4.10A)
Fasciolaria (Cinctura) rhomboidea (D,Y)
Terebraspira sparrowi (D,Y) (Figure 4.12F)
Triplofusus duplinensis (D,Y) (Figure 4.12C)
Heilprinia carolinensis (D)
Hesperisternia filicata (D,Y)
Fulguropsis excavatum (D)
Sinistrofulgur contrarium (D,Y)
Scaphella trenholmii (D)
Volutifusus obtusus (D,Y)
Volutifusus spengleri (D,Y)
Oliva carolinensis (D)
Cancellaria rotunda (D)
Extractrix hoerlei (D,Y)
Ventrilia carolinensis (D,Y)
Contraconus adversarius (D,Y) (Figure 4.10I)
Bivalvia
Costaglycymeris subovata (D,Y)
Carolinapecten eboreus (D,Y) (Figure 4.10J)
Christinapecten decemnarius (D,Y)
Argopecten comparilis (D,Y)
Euvola hemicyclicus (D,Y) (also found in the contemporaneous Ochopee Member, Figure 4.15B)
Chione cribraria (D,Y)
Mercenaria carolinensis (D)

FIGURE 4.10 Index fossils for the Pinecrest Member, Tamiami Formation. A = *Latecphora bradleyae* (Petuch, 1988), length 68 mm; B = *Melongena taurus* Petuch, 1994, length 210 mm; C = *Siphocypraea trippeana* Parodiz, 1988, length 57 mm; D = *Pyruella rugosicostata* Petuch, 1982, length 65 mm; E = *Hystrivasum locklini* (Olsson and Harbison, 1953), length 96 mm; F = *Akleistostoma floridana* (Mansfield, 1931), length 66 mm; G = *Cymatophos lindae* Petuch, 1991, length 69 mm; H = *Pterorhytis fluviana* (Dall, 1903), length 53 mm; I = *Contraconus adversarius* (Conrad, 1840), length 98 mm; J = *Carolinapecten eboreus* (Conrad, 1833), length 86 mm; K = *Hexaplex hertweckorum* (Petuch, 1988), length 80 mm.

FIGURE 4.11 Index fossils for the Pinecrest Member, Tamiami Formation. A = *Lithoconus druidi* (Olsson, 1967), length 147 mm; B = *Siphocypraea (Pahayokea) mansfieldi* Petuch, 1998, length 76 mm (listed as an *Akleistostoma* species by Petuch, 2004: 174); C = *Phyllonotus globosus* (Emmons, 1858), length 79 mm; D = *Spuriconus cherokus* (Olsson and Petit, 1964), length 73 mm; E = *Tropochasca lindae* Petuch, 1991, length 107 mm; F = *Pseudodusta hertweckorum* (Petuch, 1991), length 57 mm; G = *Vermicularia recta* (Olsson and Harbison, 1953), clump length 226 mm; H = *Eichwaldiella magnasulcus* Petuch, 1991, length 113 mm; I = *Strombus floridanus* Mansfield, 1930, length 66 mm; J = *Calusacypraea (Myakkacypraea) briani* (Petuch, 1996), length 74 mm; K = *Eichwaldiella pontoni* (Mansfield, 1931), length 165 mm; L = *Nodipecten collierensis* (Mansfield, 1931), length 92 mm.

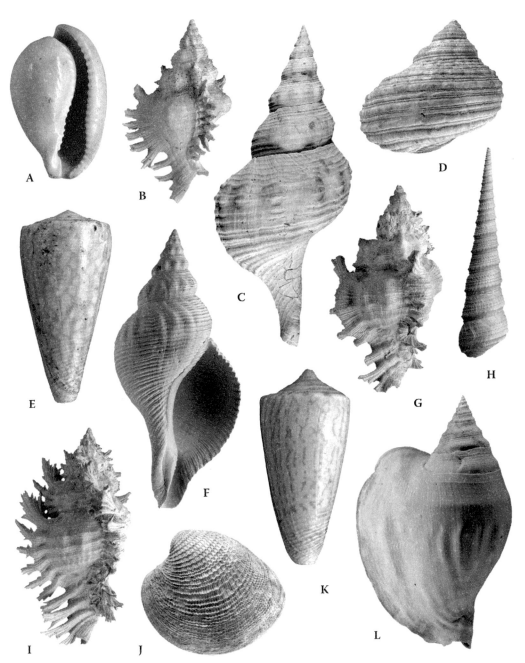

FIGURE 4.12 Index fossils for the Pinecrest Member, Tamiami Formation. A = *Calusacypraea globulina* Petuch, 2004, length 55 mm; B = *Chicoreus xestos* E. Vokes, 1974, length 38 mm; C = *Triplofusus duplinensis* (B. Smith, 1940), length 356 mm; D = *Turbo (Taenioturbo) lindae* Petuch, 1994, width 32 mm; E = *Eugeniconus paranobilis* (Petuch, 1991), length 36 mm; F = *Terebraspira sparrowi* (Emmons, 1858), length 148 mm; G = *Chicoreus stephensae* Petuch, 1994, length 42 mm; H = *Apicula gladeensis* (Mansfield, 1931), length 103 mm; I = *Chicoreus floridanus* E.Vokes, 1965, length 58 mm; J = *Chione procancellata* Mansfield, 1932, length 54 mm; K = *Dauciconus bassi* (Petuch, 1991), length 35 mm; L = *Macrostrombus hertweckorum* (Petuch, 1991), length 176 mm.

FIGURE 4.13 View of a bioherm of the worm gastropod *Vermicularia recta*, intercalated between two layers of the oyster *Hyotissa meridionalis*. This assemblage is found in Unit 9 of the Pinecrest Member. A similar, but larger and thicker, bed of worm gastropods occurs above the massive oyster layer (Unit 8).

Macrocallista reposta (D,Y)
Astarte concentrica (D,Y)

Kissimmee Facies of the Pinecrest Member Endemics
Gastropoda
Akleistostoma transitoria (Figure 4.14G)
Akleistostoma hughesi (Figure 4.14A) (previously thought (Petuch, 2004: 194) to have come
    from the Kissimmee Facies of the younger Fruitville Member, but now known to be from
    the older Pinecrest equivalent)
Chicoreus shirleyae (Figure 4.14D)
Pterorhytis lindae (Figure 4.14K)
Trossulasalpinx kissimmeensis (Figure 4.14J)
Terebraspira kissimmeensis
Terebraspira lindae (Figure 4.14H)
Terebraspira osceolai (Figure 4.14B)
Pyruella basingerensis
Hystrivasum olssoni (Figure 4.14C) (also found, rarely, at the Bird Road excavations in
    Miami, in the lower beds of the Golden Gate Member)
Hystrivasum shrinerae
Hystrivasum vokesae (Figure 4.14E)
Gradiconus duerri (Figure 4.14I)
Jaspidiconus jaclynae

FIGURE 4.14 Index fossils for the Kissimmee Facies of the Pinecrest Member of the Tamiami Formation. A = *Akleistostoma hughesi* (Olsson and Petit, 1964), length 48 mm; B = *Terebraspira osceolai* Petuch, 1994, length 98 mm; C = *Hystrivasum olssoni* (E.Vokes, 1966), length 107 mm; D = *Chicoreus shirleyae* E.Vokes, 1974, length 80 mm; E = *Hystrivasum vokesae* (Hollister, 1971), length 75 mm; F = *Carolinapecten darling-tonensis* (Dall, 1898), length 178 mm; G = *Akleistostoma transitoria* (Olsson and Petit, 1964), length 70 mm; H = *Terebraspira lindae* Petuch, 1994, length 142 mm; I = *Gradiconus duerri* (Petuch, 1994), length 62 mm; J = *Trossulasalpinx kissimmeensis* Petuch, 1994, length 28 mm; K = *Pterorhytis lindae* Petuch, 1994, length 47 mm.

Bivalvia

Carolinapecten darlingtonensis (Figure 4.14F) (also shared with the Duplin Formation; its
  only occurrence in Florida is in the Kissimmee Facies)
Cardita seminolensis
Pleuromeris pitysia
Carditamera dasytes

The marine communities of the Pinecrest Member were described by the senior author (Petuch,
2004: 171–177) and include the *Strombus floridanus* Community (shallow muddy sand areas and
intertidal mud flats), the *Apicula gladeensis* Community (shallow lagoonal turritellid beds), the
Nodipecten floridensis Community (deep lagoonal pectinid beds), and the *Hyotissa meridionalis*
Community (shallow lagoon oyster beds).

THE OCHOPEE MEMBER, TAMIAMI FORMATION

History of Discovery

When the Tamiami Trail (U.S. Route 41) was being constructed in the late 1920s and early 1930s,
a hard, distinctive limestone was exposed in shallow excavations on the southwestern side of the
Everglades. This was later named the Tamiami Limestone by Mansfield (1939). The Tamiami
material was extensively excavated, crushed, and used as the bed for the new trans-Florida highway.
Because of the abundance of exposures, Mansfield was able to study the fossil fauna in depth,
describing several new species.

After more research was undertaken in the 1950s and 1960s (by Parker, Puri, Vernon, and Olsson;
all cited previously in Chapter 1), it was found that Mansfield's "Tamiami Limestone" was simply
a facies of a much larger, lithologically complex unit. Preserving the well-known name "Tamiami,"
Hunter (1968: 443) expanded the formational concept to include the other contemporaneous lithologic units around the Everglades area, naming the entire suite the "Tamiami Formation." She then
renamed Mansfield's original "Tamiami Limestone" as the "Ochopee Limestone Member." Although
this was not in complete concordance with the Code of Stratigraphic Nomenclature, Hunter's scheme
has been accepted by consensus, and we follow her nomenclature in this book, referring to this unit
as the Ochopee Member of the Tamiami Formation.

Lithologic Description and Areal Extent

From Mansfield (1939: 8; condensed here):

> The matrix of the Tamiami limestone [authors' note: the Ochopee Member] consists mainly of a dirty-
> white to gray, rather hard, porous, nonoölitic limestone with inclusions of clear quartz sand. The faunas,
> so far studied, include 6 genera of gastropods, 15 genera of pelecypods, and 2 genera of echinoids ...
> Among the pelecypods, the scallops and oysters are the most conspicuous forms, both in numbers of
> species and individuals and in the rather large size that some of them attained. The echinoid *Encope
> macrophora tamiamiensis* Mansfield was found at three localities ... .

Hunter's description of the Ochopee Limestone Member essentially follows Mansfield's:
" ... Light grey to white, hard, sandy limestone (a calcarenite) containing abundant identifiable
mollusk molds and well preserved pectens, oysters, barnacles, and echinoids."

The Ochopee Member is confined to the southwestern Everglades area (Figure 4.8), in southern
and western Collier, extreme southern Lee, northern Monroe, and western Dade Counties. The area
corresponds to a large, shallow carbonate bank that was deposited along and behind the Immokalee
Reef Tract, the southern end of the Immokalee Archipelago, and the southern part of the Hendry
Platform. Carbonate fines eroding off this actively growing coral reef area accumulated in a large,

fanlike structure and interfingered with the quartz sand banks of the central and northern Hendry Platform (the Pinecrest Member). Quartz sand also eroded off the Immokalee Archipelago and this, combined with the central Hendry Platform sands, contributed to the siliciclastic fraction seen in the typical Ochopee limestone. The "Bonita Springs Marl Member" of Missimer (1992) is here considered to be a geographically restricted, unconsolidated dolosilt facies of the Ochopee.

The Ochopee Member varies in thickness, from 10 m in northwestern Collier County, to over 20 m at Mule Pen Quarry, East Naples, Collier County, to over 15 m at the stratotype at Ochopee (Missimer, 1992). In the stratotype area of southeastern Collier, extreme northwestern Dade, and northernmost Monroe Counties, the Ochopee Member lies nearly at the surface, at 1–3 m depth. From there, the Ochopee dips to the west and south, occurring at depths below 10–20 m at the coast. These surface contours correspond to the geomorphology of the Pliocene carbonate bank. Several workers, such as Missimer (1992), have shown the Ochopee as being subjacent to the Pinecrest, and this has been adopted by many researchers and government offices. The presence of many key Pinecrest index fossils in the Ochopee (listed and illustrated in this section), however, readily demonstrates that the two members are contemporaneous, interfingering units. Workers such as Missimer, using well cores, have misidentified older Buckingham indurated beds as being the Ochopee, and this mistake has been carried on in many subsequent papers and technical reports.

Because the member is so close to the surface near the stratotype area, the limestones were subject to extensive, long-term groundwater infiltration and leaching. Because of this, all aragonitic fossils have been dissolved away and remain only as casts and molds (often identifiable). Calcitic fossils, however, remained intact and are often very well preserved.

Stratotype

From Hunter (1968: 445; after Mansfield, 1939): " … rock pits [exist] between Carnestown (intersection of State Route 29 and U.S. Route 41) and the village of Ochopee, on U.S. Route 41, Collier County." As a point of interest, Ochopee became slightly famous for (according to its post office) as having the smallest such facility in the U.S.

Age and Correlative Units

Based on the presence of key stratigraphic index fossils, such as *Euvola hemicyclicus*, *Nodipecten collierensis*, and *Chesapecten madisonius carolinensis*, the Ochopee Member is now known to date from the early middle Piacenzian Pliocene. The member is exactly correlative with the lower beds of the Mogarts Beach Member of the Yorktown Formation, Chesapeake Group of Virginia and northern North Carolina, the middle beds of the Duplin Formation of southern North Carolina and northern South Carolina, the upper beds of the Goose Creek Formation of South Carolina, and the lower beds of the *Cancellaria* Zone of the Jackson Bluff Formation of the Florida Panhandle.

Ochopee Index Fossils

Only calcitic fossils are preserved in the Ochopee, and these are mostly bivalves. Although containing some widespread mid-Piacenzian species, the Ochopee does contain a few species that have been found only in the stratotype area. This endemism demonstrates that the carbonate bank along the southern edge of the Hendry Platform must have had its own unique set of oceanographic parameters. This area supported immense aggregations of scallops belonging to several genera, including *Argopecten*, *Carolinapecten*, *Christinapecten*, *Chesapecten*, *Lindapecten*, *Nodipecten*, *Euvola*, and *Amusium*. The Ochopee Member is also the only Piacenzian-aged unit of the Everglades area to contain scallops of the genus *Chesapecten* (*C. madisonius carolinensis*). The genus has never been found in any facies of the contemporaneous Pinecrest Member and must have been ecologically restricted to the southern Hendry Platform. Some of the more important Ochopee index fossils include:

Endemic Ochopee species (restricted to the stratotype area)
Gastropoda
Amaea (Ferminoscala) boylae
Cirsotrema woodringi
Bivalvia
Argopecten evergladesensis (Figure 4.15E)
Carolinapecten gladensis (Figure 4.15F)
Christinapecten tamiamiensis

Species shared with the Duplin and Yorktown Formations
Bivalvia
Amusium mortoni
Chesapecten madisonius carolinensis (Figure 4.15G)
Euvola hemicyclicus (Figure 4.15B)

Species shared with the Pinecrest Member
Bivalvia
Nodipecten collierensis (Figure 4.15A)
Hyotissa meridionalis (Figure 4.15D) (and form *monroensis*)
Echinodermata–Echinoidea
Encope tamiamiensis (Figure 4.15C)

THE GOLDEN GATE MEMBER, TAMIAMI FORMATION

History of Discovery

During his doctoral dissertation research in the late 1970s, J.F. "Jack" Meeder discovered richly fossiliferous limestones and unconsolidated calcarenites that were being uncovered in housing developments, canal digs, and quarries in Collier County. Upon closer examination, Meeder found that these beds represented zonated reef tracts and that they contained an unprecedented number of coral species, the richest ever found in the U.S. This important discovery was published in 1979, as a field guide for a Miami Geological Society field trip to Collier County. In that work, Meeder illustrated several of the more prominent coral species, mapped the approximate extent of the reef systems, and discussed and illustrated the zonational patterns. A year later (1980), he published the first formal paper on the reef systems, describing the lithology of the limestones and associated carbonate facies, and discussing the reefal depositional environments. A complete survey of the reef assemblages of Collier and Lee Counties, including mapping, lithology, and faunal identification, was compiled by Meeder in his dissertation (1987, unpublished).

In 1985, while teaching at Florida International University in Miami, the senior author was notified of large local construction excavations that were bringing up beautifully preserved fossil corals and mollusks. These were prominent in spoil piles heaped around artificial lakes being dug for the Lakes of the Meadows housing development west of Miami on Bird Road. After several months of collecting and research, it became apparent that the Bird Road coral reef and molluscan assemblages were very similar to those found at several of Meeder's research sites in Collier County, such as at the Golden Gate housing development excavations and at the Mule Pen Quarry (later Florida Rock Industries, Inc., Naples Quarry). When comparing Meeder's discoveries in Collier County with the faunal and lithological data from Bird Road, the first indications of the Pliocene Pseudoatoll came to light. The senior author published these data the following year (Petuch, 1986), incorporating Meeder's research and that of the U.S. Geological Survey (Water Resources Division). In that study (Swayze and Miller, 1984), U.S. Geological Survey researchers showed that the reefs under Miami were actually part of a larger continuum of coral assemblages that extend all the way to northern Palm Beach County (the Zone of Secondary Permeability, a vuggy reefal limestone containing water).

FIGURE 4.15 Index fossils for the Ochopee Member of the Tamiami Formation. A = *Nodipecten collierensis* (Mansfield, 1931), length 96 mm; B = *Euvola hemicyclicus* (Ravenel, 1834), length 48 mm; C = *Encope tamiamiensis* Mansfield, 1931, length 67 mm; D = *Hyotissa meridionalis* (Heilprin, 1886), length 92 mm; E = *Argopecten evergladesensis* (Mansfield, 1931), length 32 mm; F = *Carolinapecten gladensis* (Mansfield, 1936), length 89 mm; G = *Chesapecten madisonius carolinensis* (Conrad, 1873), length 210 mm (specimen with 13 ribs).

In his overview of the Tamiami Formation, Missimer (1992: 67) resurrected the term "Golden Gate Reef" from Meeder's unpublished dissertation and elevated it to member status. We have shortened Missimer's informal name and referred to the reefal limestones that surround the Everglades basin as the Golden Gate Member.

Lithologic Description and Areal Extent

Building on the descriptions of Meeder and Missimer, we here define the Golden Gate Member as an interbedded set of lithofacies dominated by limestones and unconsolidated calcilutites. These include vuggy moldic limestone with high primary porosity, highly fossiliferous limestone containing densely packed assemblages of well-preserved aragonitic corals (biomicrudites), layers of unconsolidated calcilutites and dolosilts packed with perfectly preserved mollusks and corals, and layers of spar-cemented recrystallized (calcitic) corals. The limestones vary in color from light gray to pale tan or cream. The unconsolidated calcilutites, dolosilts, and fine-grained calcarenites are colored pale tan, light gray, or pale blue-gray. Corals and mollusk shells preserved within these intercalated unconsolidated units are colored either pale tan or blue-gray, weathering to white on exposure. Small amounts of quartz sand are present in some of the limestones and calcilutites, rarely more than 15%. Some limestones, particularly those at the base of the member, are dolomitized and devoid of fossils. In some facies, thin stringers of quartz sand are present, usually at localities that were adjacent to Immokalee Island.

The areal distribution of the Golden Gate Member conforms to the shape of the reef tracts of the Everglades Pseudoatoll (Figure 4.8 and Figure 4.21). The member is essentially "U"-shaped, following the outline of the main reef structures, from southern Lee, Collier, Monroe, Dade, Broward, and Palm Beach Counties. This reefal limestone member varies in thickness from 15 m at the Mule Pen Quarry, East Naples, Collier County, to 5 m in southern Collier County (Missimer, 1992: Core Log 11), to over 50 m under the Atlantic Coastal Ridge (Swayze and Miller). At the Mule Pen Quarry, the reefal limestones are 5 m below suface, whereas under eastern Palm Beach County (adjacent to the Florida Turnpike) they averaged 10 m below surface.

Stratotype

Wishing to preserve Meeder's original name (used extensively by Missimer), we here designate the type locality of the Golden Gate Member as the old Mule Pen Quarry (most recently, Florida Rock Industries, Inc., Naples Quarry) on Immokalee Road, East Naples, Collier County, near the Golden Gate housing subdivision (Sec. 23, T. 48 S., R. 26 E., Collier County). The quarry is now flooded as a large lake (over $^3/_4$ mile across) surrounded by a housing development. The member is exposed at a depth of 5 m below lake surface and can be accessed by scuba diving.

Age and Correlative Units

The basal beds of the Golden Gate Member, encountered at Bird Road, contained many classic Duplin, Yorktown, and Pinecrest index fossils, demonstrating that the reefal deposition started in the early middle Piacenzian Pliocene. Some of the widespread Duplin–Yorktown guide fossils found at Bird Road (Petuch, 1994) included the gastropods *Sconsia hodgei*, *Extractrix hoerlei*, *Ptychosalpinx multirugata*, *Contraconus adversarius*, and *Diodora (Glyphis) redemicula*. Only the upper beds of the Golden Gate were excavated during the last 10 years of excavations at the Mule Pen Quarry, Collier County, exposing the chronological equivalent of the Fruitville Member (described next). Some of these younger index fossils, also found in the upper beds of the Yorktown and Duplin Formations, include the gastropod *Calliostoma conradianum* and the bivalve *Chama emmonsi*. The lower Golden Gate beds in Collier County, those equivalent to the Pinecrest and Ochopee Members, were shown by Missimer (1992: Core Log 11).

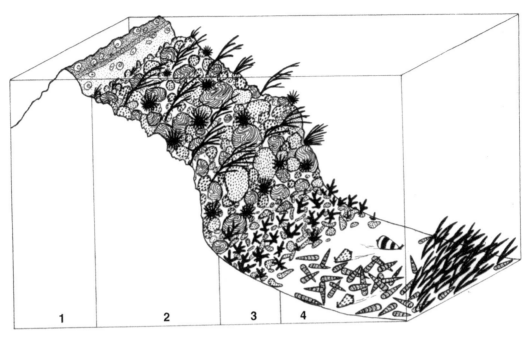

FIGURE 4.16 Schematic diagram showing a cross-section of a typical coral reef system of the Everglades Pseudoatoll during the late Pliocene (upper Golden Gate Member time). The ecological zones and their associated communities include: Zone 1 = Reef Crest (*Pocillopora crassoramosa* Community); Zone 2 = Reef Platform (*Dichocoenia tuberosa* Community, front section, and *Isophyllia desotoensis* Community, back section); Zone 3 = Back Reef (*Stylophora affinis* Community); and Zone 4 = Reef Lagoon (*Antillia bilobata* Community).

The Golden Gate Member is now known to span the middle and late Piacenzian Pliocene, with the greatest development occurring during the late Piacenzian (Fruitville equivalent time). The member is exactly correlative with the Mogarts Beach and Moore House Members of the Yorktown Formation and the Edenhouse Member of the Chowan River Formation, all of Virginia and northern North Carolina, the Duplin Formation of southern North Carolina and northern South Carolina, and the "*Cancellaria* Zone" of the Jackson Bluff Formation of the Florida Panhandle. The diachronous nature of Golden Gate deposition reflects the long-lived and continuous growth of the pseudoatoll coral reef systems.

Golden Gate Index Fossils

The primary index fossil groups for the Golden Gate Member are the scleractinian corals, which often occur in densely packed assemblages. Because the pseudoatoll corals were zonated by wave energy, the species were segregated into distinct bands of ecosystems, often widely separated. Generally, four broad energy-related zones can be seen across most of the reef tracts. These include (from open ocean towards shore): the **Reef Crest** (high-energy surf environment); the **Reef Platform** (lower energy surge environment, often subdivided into a front reef platform and a back reef platform); the **Back Reef** (low-energy environment); and the **Reef Lagoon** (quiet, deeper water environment). Collecting at several sites at the Bird Road lake excavations, on an east–west transect with each site separated by 0.5 km, the senior author (Petuch, 1986) was able to demonstrate the existence of this zonational pattern on the Miami Reef Tract of the Everglades Pseudoatoll (shown here on Figure 4.16).

The Pliocene reefs at Mule Pen Quarry, the Golden Gate area, and at Bird Road contain the richest scleractinian coral fauna ever found in the Neogene western Atlantic, with over 100 species having been collected. The senior author (Petuch, 2004: 178) originally stated that the Golden Gate Member contained over 70 species, but subsequent analyses of Mule Pen Quarry material have shown that the number actually exceeds 100, and may reach 120. This species richness is unprecedented in the U.S. and approximates that of the Recent Moluccas Islands of Indonesia (the most diverse living coral fauna). Indeed, many of the coral genera found at Mule Pen are now extinct in the Atlantic Ocean but are flourishing in the South Pacific. At least half the Golden Gate coral species were originally described from the Pliocene reefs of the Dominican Republic (Vaughan, 1919; Petuch, 2004: 176–185), demonstrating a strong Pliocene West Indian–Caribbean faunal influence. Some of the more abundant and indicative of the Golden Gate coral species include (arranged by zone; D, also found in the Pliocene of the Domincan Republic, C, also found in the Pliocene of Cuba):

Cnidaria–Scleractinia
Reef Crest
Pocillopora crassoramosa (Figure 4.18E) (D)
Pocillopora barracoaensis (C)
Goniopora jacobiana (Figure 4.18J) (D)
Astrocoenia meinzeri (Figure 4.17D) (D)
Reef Platform (front region)
Dichocoenia cf. *eminens*
Dichocoenia tuberosa (Figure 4.17B) (D)
Barysmilia intermedia (Figure 4.17C) (D)
Stephanocoenia spongiformis (D)
Montastrea brevis (Figure 4.17H) (D)
Montastrea endothecata (Figure 4.18G) (D)
Montastrea limbata (D)
Solenastrea globosa (D)
Siderastrea dalli (Figure 4.18I) (also found in Pinecrest coral bioherms)
Siderastrea pliocenica (Figure 4.18I) (D) (also found in Pinecrest coral bioherms)
Reef Platform (back region)
Isophyllia desotoensis (Figure 4.17E)
Isophyllia cf. *sinuosa* (Figure 4.18A)
Mussa affinis (D)
Syzygophyllia dentata (Figure 4.17A) (D)
Asterosmilia exarata (D)
Diploria sarasotana (Figure 4.18B) (also found in Pinecrest coral bioherms)
Thysanus excentricus (Figure 4.18D) (D) (also found in Pinecrest coral bioherms)
Agaricia dominicensis (D)
Back Reef
Stylophora affinis (Figure 4.17F) (D)
Stylophora granulata (Figure 4.18H) (D)
Stylophora minor (D) (also found in Pinecrest coral bioherms)
Porites barracoaensis (C)
Porites matanzasensis (C)
Acropora panamensis subspecies (D) (Figure 4.18C)
Oculina sarasotaensis (also found in Pinecrest coral bioherms)
Thysanus corbicula (Figure 4.17I) (D)
Septastrea matsoni (D)
Reef Lagoon
Antillia bilobata (Figure 4.17J) (D)
Antillia dubia (Figure 4.18F) (D)

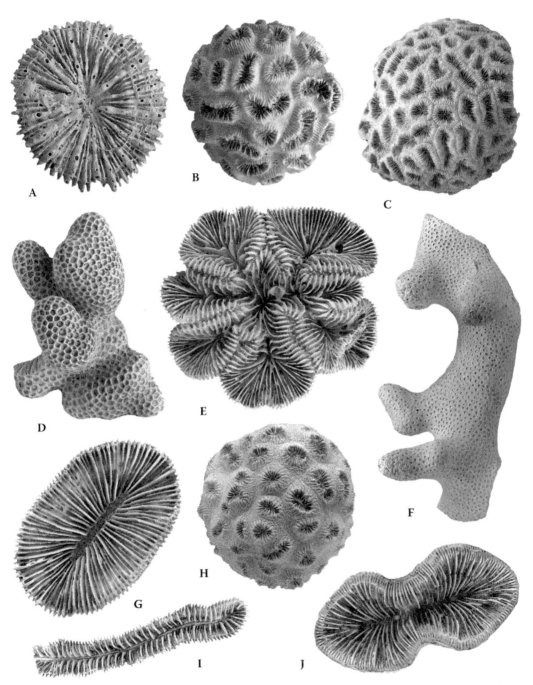

FIGURE 4.17 Index fossils for the Golden Gate Member of the Tamiami Formation. A = *Syzygophyllia dentata* (Duncan, 1863), length 43 mm; B = *Dichocoenia tuberosa* Duncan, 1863, length 63 mm; C = *Barysmilia intermedia* Duncan, 1863, length 61 mm; D = *Astrocoenia meinzeri* Vaughan, 1919, length 55 mm; E = *Isophyllia desotoensis* Weisbord, 1974, length 89 mm; F = *Stylophora affinis* Duncan, 1863, length 121 mm; G = *Antillia walli* Duncan, 1863, length 47 mm; H = *Montastrea brevis* (Duncan, 1863), length 58 mm; I = *Thysanus corbicula* Duncan, 1863, length 66 mm; J = *Antillia bilobata* Duncan, 1863, length 75 mm.

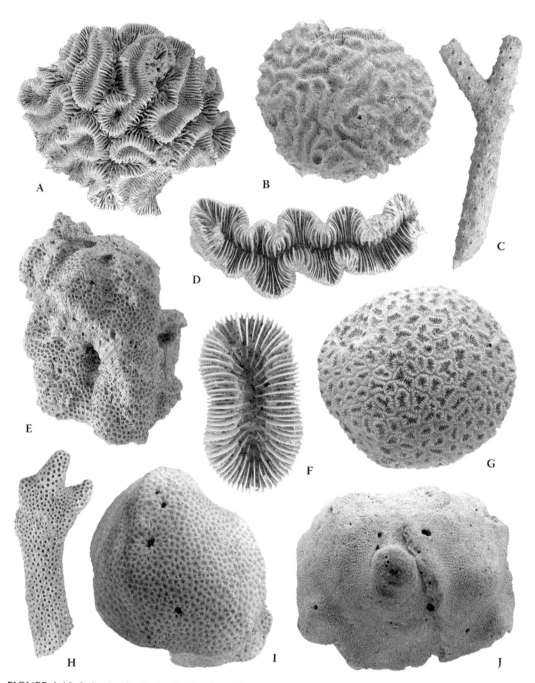

FIGURE 4.18 Index fossils for the Golden Gate Member of the Tamiami Formation. A = *Isophyllia* cf. *sinuosa* (Ellis and Solander, 1786), length 87 mm; B = *Diploria sarasotana* Weisbord, 1974, length 104 mm (specimen with *Ceratoconcha prefloridana* (Brooks and Ross, 1960) coral-dwelling barnacles and polychaete worm tubes); C = *Acropora* cf. *panamensis* Vaughan, 1919, length 74 mm; D = *Thysanus excentricus* Duncan, 1863, length 114 mm; E = *Pocillopora crassoramosa* Duncan, 1863, length 115 mm; F = *Antillia dubia* (Duncan, 1863), length 45 mm; G = *Montastrea endothecata* (Duncan, 1863), length 128 mm; H = *Stylophora granulata* Duncan, 1863, length 87 mm; I = *Siderastrea pliocenica* Vaughan, 1919, length 119 mm (specimen with apertural tube openings of encased *Magilus streami* Petuch, 1994; J = *Goniopora jacobiana* Vaughan, 1919, length 124 mm.

Antillia walli (Figure 4.17G) (D)
Manicina pliocenica (D)
Placocyathus variabilis (D)
Placocyathus barretti (D)
Flabellum dubium (D)
Flabellum exaratum (D)

As can be seen from these scleractinian coral assemblages, the Reef Crest is dominated by low, massive, encrusting species. These share the habitat with coralline algae of the genera *Goniolithon* and *Lithothamnion*, and massive ramose forms of the hydrocoral *Millepora* (Petuch, 2004: plate 58). The front part of the Reef Platform is dominated by massive (some over 2 m in diameter), rounded heads of various types of star corals. The back part of the Reef Platform, in quieter water conditions, is dominated by various types of mushroom, lettuce, and brain corals. The Back Reef is dominated by solitary cup corals and rose corals. These share the habitat with Turtle Grass (*Thalassia*) beds and open areas covered with beds of various types of turritellid gastropods.

Besides the corals, the mollusks also make up a large portion of the carbonate bioclast component, especially in the unconsolidated calcilutite and dolosilt beds. As in the corals, the molluscan fauna falls into two broad categories: a large component of endemic Floridian species and a smaller component of West Indian–Caribbean species, previously known only from the Pliocene of the Dominican Republic or Jamaica. Like the West Indian corals, these mollusks are only found on or near the pseudoatoll reefs and have not been found anywhere else in Florida. Some of the more abundant and indicative species include:

Species endemic to the Pseudoatoll Reefs (Golden Gate restricted)
Gastropoda
Astraea (Lithopoma) tectariaeformis
Cerithioclava turriculus
Cyphoma finkli
Cyphoma miamiensis
Cyphoma viaavensis (Figure 4.10C)
Pseudocyphoma carolae (Figure 4.19I)
Jenneria violetae (Figure 4.20C) (also found on Pinecrest coral bioherms)
Decoriatrivia miccosukee
Pseudadusta ketteri (Figure 4.20J) (also found on Fruitville coral bioherms)
Pseudadusta lindae (Figure 4.19D)
Siphocypraea grovesi (Figure 4.20I)
Magilus streami (Figure 4.19F) (also found on Pinecrest coral bioherms)
Babelomurex lindae (Figure 4.19A) (also found on Pinecrest coral bioherms)
Subpterynotus miamiensis
Trossulasalpinx vokesae
Latirus miamiensis (Figure 4.19E)
Pleuroploca lindae (Figure 4.19G)
Busycoarctum superbus (Figure 4.20E)
Fulguropsis radula (Figure 4.20A)
Lindafulgur miamiensis (Figure 4.20B)
Hesperisternia dadeensis
Hesperisternia joelshugari
Hesperisternia miamiensis
Parametaria lindae
Solenosteira mulepenensis (Figure 4.20F)
Mitra (Scabricola) lindae (Figure 4.19H) (also found on Pinecrest coral bioherms)

FIGURE 4.19 Index fossils for the Golden Gate Member of the Tamiami Formation. A = *Babelomurex lindae* Petuch, 1988, length 15 mm; B = *Pachycrommium guppyi* (Gabb, 1873), length 23 mm; C = *Cyphoma viaavensis* Petuch, 1986, length 26 mm; D = *Pseudadusta lindae* (Petuch, 1986), length 37 mm; E = *Latirus miamiensis* Petuch, 1986, length 45 mm; F = *Magilus streami* Petuch, 1994, length 21 mm; G = *Pleuroploca lindae* Petuch, 2004, length 87 mm; H = *Mitra (Scabricola) lindae* Petuch, 1986, length 26 mm; I = *Pseudocyphoma carolae* (Petuch, 1986), length 20 mm; J = *Eugeniconus irisae* Petuch, 2004, length 33 mm; K = *Petricoxenica concoralla* (H. Vokes, 1976), length 44 mm; L = *Virgiconus miamiensis* (Petuch, 1986), length 35 mm.

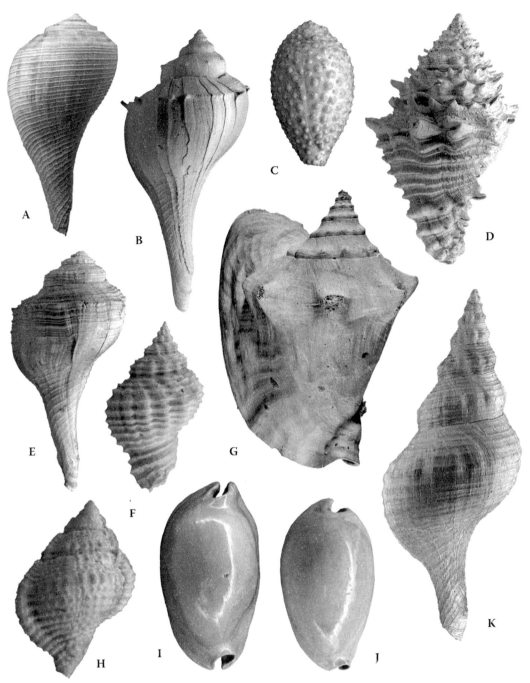

FIGURE 4.20 Index fossils for the Golden Gate Member of the Tamiami Formation. A = *Fulguropsis radula* Petuch, 1994, length 98 mm; B = *Lindafulgur miamiensis* (Petuch, 1991), length 152 mm; C = *Jenneria violetae* Petuch, 1991, length 23 mm; D = *Hystrivasum chilesi* Petuch, 1994, length 126 mm; E = *Busycoarctum superbus* (Petuch, 1994), length 147 mm; F = *Solenosteira mulepenensis* Petuch, 1994, length 32 mm; G = *Eustrombus dominator* (Pilsbry and Johnson, 1917), length 231 mm; H = *Bursa (Marsupina) proavus* Pilsbry, 1922, length 24 mm; I = *Siphocypraea grovesi* Petuch, 1998, length 58 mm; J = *Pseudadusta ketteri* (Petuch, 1994), length 54 mm; K = *Triplofusus harveyensis* (Mansfield, 1930), length 287 mm (also found in the "*Cancellaria* Zone" of the Jackson Bluff Formation).

Jaspidiconus hertwecki
Contraconus schmidti
Eugeniconus irisae (Figure 4.19J)
Spuriconus martinshugari
Virgiconus miamiensis (also found on Pinecrest coral bioherms)
Bivalvia
Petricoxenica concoralla (Figure 4.19K)

Species shared with the Dominican Republic and Jamaica
Acmaea actina
Bayericerithium obesum
Cerithium dominicensis
Cerithium turriculum
Modulus basileus
Hipponix ceras
Bursa (Lampasopsis) amphitrites
Bursa (Marsupina) proavus (Figure 4.20H)
Eustrombus dominator (Figure 4.20G)
Pachycrommium guppyi (Figure 4.19B)
Coralliophila miocenica
Columbella submercatoria

The marine communities of the Golden Gate Member were described by the senior author (Petuch, 2004: 176–185) and include the *Pocillopora crassoramosa* Community (Reef Crest), the *Dichocoenia tuberosa* Community (Reef Platform, front section), the *Isophyllia desotoensis* Community (Reef Platform, back section), the *Stylophora affinis* Community (Back Reef), and the *Antillia bilobata* Community (Reef Lagoon).

THE FRUITVILLE MEMBER, TAMIAMI FORMATION

History of Discovery

Taking into account the lack of formal descriptions for most of the Tamiami members, Waldrop and Wilson (1990) proposed a new formational name for the Pliocene sequence at Sarasota, the "Fruitville Formation" (named for Fruitville, a small town east of the city of Sarasota). All the shell beds above Unit 11 at the APAC pit (Figure 4.3) were included in their new formation, and the name was presented as a replacement for the loosely described Pinecrest Member. Upon closer scrutiny, it can be seen that Waldrop and Wilson took only a biostratigraphic approach and created a biozone and not a true lithostratigraphic unit. Because of this, we feel that the Fruitville should not be afforded full formational status.

In light of better lithologic descriptions of the Pinecrest Member (Missimer, 1992; earlier in this chapter), we here remove Units 5–9 of the APAC section from Waldrop and Wilson's Fruitville and include these units in our expanded definition of the Pinecrest. Because several workers (such as Emily Vokes in her later muricid studies) have used the name "Fruitville Formation" for the Sarasota beds, we feel it is best to preserve and redefine the name. In this book, we restrict Waldrop and Wilson's name to the upper Tamiami beds, corresponding to Unit 2, Unit 3, and Unit 4 in the APAC pit (Figure 4.3). These will be referred to as the Fruitville Member.

Lithologic Description and Areal Extent

The Fruitville Member was deposited in two distinct pulses, corresponding to two eustatic highs during the late Pliocene (Figure 4.2). Each of these marine transgressions were separated by at least

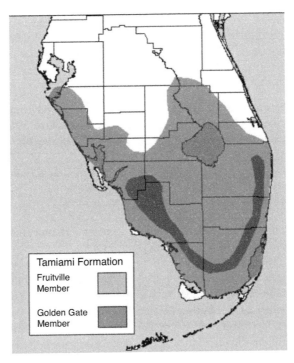

FIGURE 4.21 Areal distribution of the Fruitville and Golden Gate (upper beds) Members of the Tamiami Formation.

200,000 years of dry, terrestrial environments, producing obvious unconformities. This sequence is best seen in the APAC pit at Sarasota (Figure 4.3), which can be used as a standard reference section showing subjacent and suprajacent units. The first transgression, at 3 million years B.P., had the highest sea level and deposited Unit 3 and Unit 4 and their equivalents around the state. The second transgression, at 2.6 million years B.P., had a lower sea level and deposited Unit 2 and its equivalents. When discussing the Fruitville Member in this book, we will use these numbered units to differentiate positions within the depositional sequence.

We here define the Fruitville Member as a set of interbedded units variably composed of organic-rich quartz sand, muddy quartz sand with densely packed mollusk shell bioclasts (usually bivalves), and quartz sand with molluscan bioclasts and small amounts of clay, usually sepiolite. Limestones are absent from the Fruitville Member, although the bedding plain surfaces of some units, such as Unit 3 and Unit 4 at Sarasota, have thin (less than 10 cm) indurated duricrusts composed of fossiliferous arenaceous calcarenites. The quartz sand varies in color from white, to yellow-tan, to dark brown or dark gray. In the organic-rich sand units (such as Unit 4 at Sarasota), the beds are filled with black-stained mollusks and have a high sulfide content, with small crystals of pyrite being deposited on the shells and quartz sand grains. Fruitville molluscan fossils are very well preserved and range in color from white, to light brown, to bluish-black. These shells make up the largest portion of the carbonate fraction. The Fruitville can be differentiated from the underlying Pinecrest by having thinner quartz sand units that are muddier and contain more clay and by having beds that often contain high concentrations of pyrite crystals.

The areal distribution of the Fruitville Member (Figure 4.21) is essentially the same as the underlying Pinecrest Member. Like the Pinecrest, the Fruitville also has a distinctive Kissimmee Facies, composed of muddy, organic-rich sand generally packed with crushed mollusk shells (shell hash). The Kissimmee Embayment during Fruitville time also housed a molluscan fauna that was rich in endemic species. The member varies in thickness from 5 m at Sarasota, Sarasota County, to 8 m in Lee County, to 7 m at Fort Drum, Okeechobee County (Rucks Pit). The surface contours of the Fruitville Member conform to the bathymetric contours of the Tamiami Subsea. At Sarasota, the Fruitville lies 2–3 m below land surface, whereas at Fort Drum the member lies approximately 10–12 m below land surface.

Stratotype

We here designate the section in the APAC pit, Unit 2, Unit 3, and Unit 4 (Figure 4.3), as the stratotype of the Fruitville Member. The APAC quarry is now a lake in a Sarasota County park between Richardson Road and Interstate Highway 75 (I-75), in the Fruitville area of Sarasota, Sarasota County (1 km north of Fruitville Road, Sec. 12, T. 36 S., R. 18 E.). The member lies approximately 2 m below lake surface and can be accessed by scuba diving.

Age and Correlative Units

Based on the presence of widespread molluscan chronostratigraphic index fossils such as the bivalves *Carolinapecten bertiensis*, *Stralopecten ernsestmithi*, *Nodipecten vaccamavensis*, and *Chama emmonsi* and the gastropods *Monostiolum petiti* and *Sinistrofulgur adversarium*, the Fruitville Member is now known to date from the late Piacenzian Pliocene. The member is correlative with the Moore House Member of the Yorktown Formation (Unit 3 and Unit 4 only) and the Edenhouse Member of the Chowan River Formation (Unit 2 only), both of Virginia and northern North Carolina, the lower beds of the Waccamaw Formation of southern North Carolina and northern South Carolina, the uppermost beds of the "*Cancellaria* Zone" of the Jackson Bluff Formation of the Florida Panhandle, and the upper beds of the Golden Gate Member.

Fruitville Index Fossils

Having been separated for hundreds of thousands of years of emergent, terrestrial conditions, each of the three marine depositional pulses of the Fruitville contains a completely different molluscan fauna. Among the more interesting of the Fruitville environments was that of the huge brackish water estuary of the Myakka Lagoon System (during Unit 4 time). Immense beds of *Mulinia sapotilla* bivalves lived in these estuaries, and at the mouths of rivers, forming solid-packed layers composed almost entirely of clam shells (Figure 4.22). Another interesting environment was found in cleaner sand areas, with quiet, sheltered conditions and normal salinity. Here, extensive beds of the large pearly mussel, *Perna conradiana* (Figure 4.25), formed the basis of a rich and highly endemic ecosystem (particularly during Unit 3 time). In other areas, such as in the Kissimmee Embayment during Unit 2 time, huge beds of the spiny slipper shell, *Crepidula chamnessi*, lived attached to shell rubble and Turtle Grass.

All of these units and ecosystems housed highly endemic molluscan faunas, and these can be used as stratigraphic guide fossils. We here list some of these Fruitville endemics by the numbered units at the stratotype in the APAC pit.

Restricted to Unit 4 and equivalent beds (G, also found in the upper beds of the Golden
   Gate Member)
Gastropoda
Neritina sphaerica (Figure 4.24I)
Cerithidea lindae (Figure 4.23H)
Pyrazisinus lindae (Figure 4.23G)
Pyrazisinus sarasotaensis (Figure 4.24D)
Pyrazisinus scalinus (Figure 4.23F) (G)
Strombus sarasotaensis (G)
Akleistostoma jenniferae (Figure 4.24J)
Akleistostoma macbrideae (Figure 4.24E)

FIGURE 4.22 View of the Fruitville Formation (Unit 4, estuarine facies) of the Tamiami Formation in the APAC pit, Sarasota, Sarasota County, Florida. Here, closely packed beds of the small bivalve *Mulinia sapotilla* make up almost the entire unit. Thin layers of the large pearly mussel *Perna conradiana* are intercalated between the *Mulinia* beds and were deposited through winnowing by currents in tidal channels. These types of bivalve aggregations are typical of the Myakka Lagoon System estuarine environments.

Calusacypraea tequesta (Figure 4.23J)
Calusacypraea (Myakkacypraea) myakka (Figure 4.23A)
Pseudadusta metae (Figure 4.23I)
Trossulasalpinx lindae (Figure 4.23B)
Calotrophon myakka (Figure 4.24B)
Thais (Stramonita) sarasotana
Fasciolaria (Cinctura) sarasotaensis (G)
Melongena draperi (Figure 4.24F)
Echinofulgur jonesae (Figure 4.23E) (G)
Tropochasca metae (Figure 4.24A)
Pyruella laevis (Figure 4.24K) (G)
Pyruella schmidti
Hystrivasum lindae (Figure 4.23C) (G)
Contraconus berryi (G)
Bulla sarasotaensis (Figure 4.24C)
Bivalvia
Carolinapecten watsonensis (also found in the Jackson Bluff Formation)

Restricted to Unit 3 and equivalent beds (G, also found in the upper beds of the Golden
 Gate Member)
Gastropoda
Crepidula cannoni

Akleistostoma rilkoi (Figure 4.27D)
Calusacypraea sarasotaensis (Figure 4.26C)
Calusacypraea (Myakkacypraea) kelleyi (Figure 4.27G)
Pseudadusta kalafuti (Figure 4.27H)
Siphocypraea cannoni (Figure 4.27B)
Siphocypraea (Pahayokea) parodizi (Figure 4.26I)
Chicoreus miccosukee (G)
Pterorhytis roxaneae (Figure 4.26F)
Pterorhytis squamulosa (Figure 4.26E) (G)
Hexaplex jameshoubricki (Figure 4.26B)
Echinofulgur helenae (Figure 4.26D)
Melongena cannoni
Pyruella federicoae (G)
Pyruella turbinalis (G)
Heilprinia hasta (G)
Calophos nanus (Figure 4.27C)
Hystrivasum chilesi (G)
Hystrivasum jacksonense (also found in the Jackson Bluff Formation)
Scaphella gravesae (Figure 4.26J) (G)
Scaphella maureenae
Cancellaria cannoni (G)
Spuriconus streami (Figure 4.27K)

Restricted to Unit 2 and equivalents (G, also found in the upper beds of the Golden Gate
  Member)
Gastropoda
Calusacypraea (Myakkacypraea) schnireli (Figure 4.26G)
Pseudadusta marilynae (Figure 4.27L)
Siphocypraea (Pahayokea) alligator (Figure 4.27E)
Cassis ketteri
Aspella petuchi (G)
Murexiella petuchi
Chicoreus judeae (Figure 4.27F) (G)
Trossulasalpinx maryae
Melongena sarasotaensis (G)
Dolicholatirus metae
Latirus duerri (Figure 4.26A)
Monostiolum petiti (Figure 4.26H) (also found in the Waccamaw Formation)
Hystrivasum hertweckorum (Figure 4.27)
Hystrivasum hyshugari
Cancellaria floridana (G)
Spuriconus jonesorum (G)
Spuriconus yaquensis (Figure 4.26K)
Bivalvia
Nodipecten vaccamavensis (Figure 4.27A) (also found in the Waccamaw Formation)
Stralopecten ernestsmithi (Figure 4.27J) (also found in the Waccamaw Formation)

Fruitville Kissimmee Facies (Unit 4 equivalent) (G, also found in the upper beds of the
  Golden Gate Member)
Gastropoda
Siphocypraea (Pahayokea) kissimmeensis (Figure 4.29) (G)

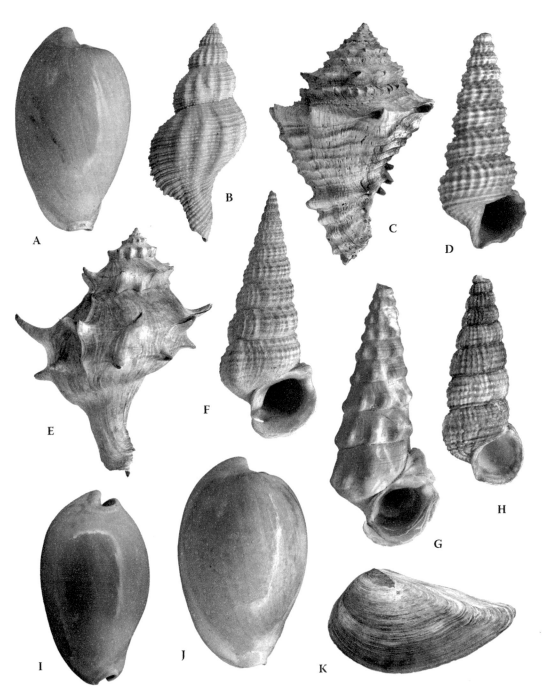

FIGURE 4.23 Index fossils for the Fruitville Member of the Tamiami Formation (Unit 4 equivalent). A = *Calusacypraea (Myakkacypraea) myakka* Petuch, 2004, length 59 mm; B = *Trossulasalpinx lindae* Petuch, 1991, length 28 mm; C = *Hystrivasum lindae* Petuch, 1994, length 84 mm; D = *Potamides matsoni* Dall, 1913, length 12 mm (also found in the Satilla beds of Georgia); E = *Echinofulgur jonesae* Petuch, 1994, length 75 mm; F = *Pyrazisinus scalinus* (Olsson, 1967), length 48 mm; G = *Pyrazisinus lindae* Petuch, 1994, length 62 mm; H = *Cerithidea lindae* Petuch, 1994, length 29 mm; I = *Pseudadusta metae* (Petuch, 1994), length 63 mm; J = *Calusacypraea tequesta* (Petuch, 1996), length 75 mm; K = *Mulinia sapotilla* Dall, 1898, length 28 mm (also found in the Satilla beds of Georgia).

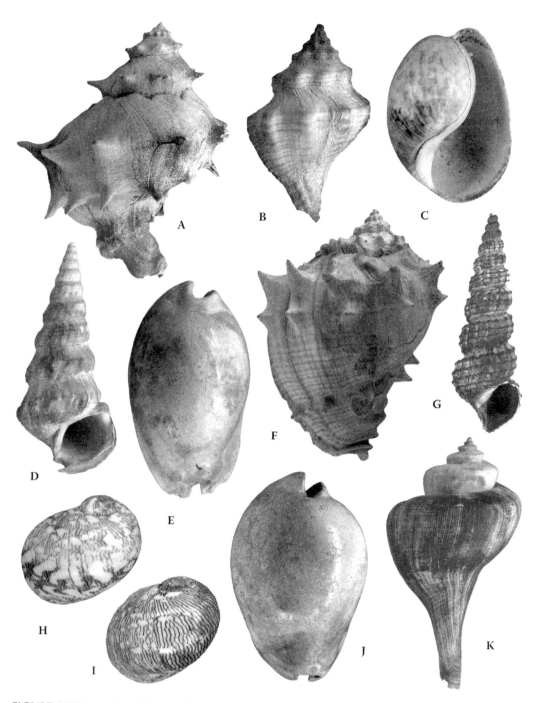

FIGURE 4.24 Index fossils for the Fruitville Member of the Tamiami Formation (Unit 4 equivalent). A = *Tropochasca metae* Petuch, 1994, length 94 mm; B = *Calotrophon myakka* Petuch, 1994, length 20 mm; C = *Bulla sarasotaensis* Petuch, 1994, length 20 mm; D = *Pyrazisinus sarasotaensis* Petuch, 1994, length 38 mm; E = *Akleistostoma macbrideae* (Petuch, 1998), length 78 mm; F = *Melongena draperi* Petuch, 1994, length 65 mm; G = *Potamides gracilior* Dall, 1913, length 16 mm; H = *Smaragdia merida* (Dall, 1903), length 5 mm (also found in the Satilla beds of Georgia); I = *Neritina sphaerica* Olsson and Harbison, 1953, length 4 mm; J = *Akleistostoma jenniferae* (Petuch, 1998), length 81 mm; K = *Pyruella laevis* Petuch, 1982, length 64 mm.

FIGURE 4.25 View of the surface of a bed of the pearly mussel *Perna conradiana* in the Fruitville Member (Unit 3 equivalent), at the Quality Aggregates pit #6, Sarasota, Sarasota County, Florida. Note the abundance of large gastropods interspersed among the crushed mussel shells, some of which include *Siphocypraea cannoni*, *Trochita floridana*, *Crepidula cannoni*, *Pleioptygma brandyceae*, and *Oliva carolinensis*. (Photo by Brian Schnirel.)

Terebraspira diegelae
Brachysycon kissimmeensis
Hystrivasum kissimmense (Figure 4.29F)
Jaspidiconus laurenae

Fruitville Kissimmee Facies (Unit 3 equivalent) (G, also found in the upper beds of the Golden Gate Member)
Gastropoda
Akleistostoma bairdi (Figure 4.28K)
Siphocypraea (Pahayokea) basingerensis (Figure 4.29J)
Littorinopsis sheaferi
Cerithidea briani (Figure 4.29K)
Cerithidea diegelae
Pyrazisinus kissimmeensis (Figure 4.29D)
Ilyanassa marthae
Terebraspira okeechobeensis (Figure 4.29B) (G)
Hystrivasum squamosum (Figure 4.29E) (G)

Fruitville Kissimmee Facies (Unit 2 equivalent) (G, also found in the upper beds of the Golden Gate Member)
Gastropoda
Akleistostoma diegelae (Figure 4.29H)

FIGURE 4.26 Index fossils for the Fruitville Member of the Tamiami Formation (Unit 2 and Unit 3 equivalents). A = *Latirus duerri* Petuch, 2004, length 52 mm (Unit 2); B = *Hexaplex jameshoubricki* Petuch, 1994, length 113 mm (Unit 3); C = *Caluscypraea sarasotaensis* (Petuch, 1994), length 41 mm (Unit 3); D = *Echinofulgur helenae* Olsson, 1967, length 98 mm (Unit 3); E = *Pterorhytis squamulosa* Petuch, 1994, length 36 mm (Unit 3); F = *Pterorhytis roxaneae* Petuch, 1994, length 38 mm (Unit e); G = *Calusacypraea (Myakkacypraea) schnireli* Petuch, 2004, length 41 mm (Unit 2); H = *Monostiolum petiti* Olsson, 1967, length 24 mm (Unit 2); I = *Siphocypraea (Pahayokea) parodizi* Petuch, 1994, length 57 mm (Unit 3); J = *Scaphella gravesae* Petuch, 1994, length 153 mm (Unit 3); K = *Spuriconus yaquensis* (Gabb, 1873), length 44 mm (Unit 2).

FIGURE 4.27 Index fossils for the Fruitville Member of the Tamiami Formation (Unit 2 and Unit 3 equivalents). A = *Nodipecten vaccamavensis* (Olsson, 1914), length 74 mm (Unit 2); B = *Siphocypraea cannoni* Petuch, 1994, length 74 mm (Unit 3); C = *Calophos nanus* Petuch, 1994, length 27 mm (Unit 3); D = *Akleistostoma rilkoi* (Petuch, 1998), length 66 mm (Unit 3); E = *Siphocypraea (Pahayokea) alligator* Petuch, 1994, length 64 mm (Unit 2); F = *Chicoreus judeae* Petuch, 1994, length 59 mm (Unit 2); G = *Calusacypraea (Myakkacypraea) kelleyi* (Petuch, 1998), length 40 mm (Unit 3); H = *Pseudadusta kalafuti* (Petuch, 1994), length 82 mm (Unit 3); I = *Hystrivasum hertweckorum* Petuch, 1994, length 66 mm (Unit 2); J = *Stralopecten ernestsmithi* (Tucker, 1931), length 63 mm (Unit 2); K = *Spuriconus streami* (Petuch, 1994), length 53 mm (Unit 3); L = *Pseudadusta marilynae* (Petuch, 1994), length 42 mm (Unit 2).

FIGURE 4.28 Index fossils for the Kissimmee Facies of the Fruitville Member, Tamiami Formation. A = *Busycon auroraensis* Petuch, 1994, length 172 mm; B = *Sinistrofulgur adversarium* (Conrad, 1862), length 181 mm; C = *Carolinapecten bertiensis* (Mansfield, 1936), length 95 mm; D = *Contraconus petiti* Petuch, 2004, length 78 mm; E = *Terebraspira maryae* Petuch, 1994, length 155 mm; F = *Siphocypraea (Pahayokea) gabrielleae* Petuch, 2004, length 67 mm; G = *Busycotypus concinnum* (Conrad, 1875), length 136 mm; H = *Siphocypraea (Pahayokea) rucksorum* Petuch, 2004, length 69 mm; I = *Volutifusus typus* Conrad, 1866, length 138 mm; J = *Brachysycon canaliferum* (Conrad, 1862), length 162 mm; K = *Akleistostoma bairdi* Petuch, 2004, length 49 mm.

FIGURE 4.29 Index fossils for the Kissimmee Facies of the Fruitville Member of the Tamiami Formation. A = *Hystrivasum palmerae* (Hollister, 1971), length 76 mm; B = *Terebraspira okeechobeensis* Petuch, 1994, length 94 mm; C = *Siphocypraea (Pahayokea) kissimmeensis* Petuch, 1994, length 53 mm; D = *Pyrazisinus kissimmeensis* (Olsson, 1967), length 82 mm; E = *Hystrivasum squamosum* (Hollister, 1971), length 125 mm; F = *Hystrivasum kissimmense* (Hollister, 1971), length 135 mm; G = *Siphocypraea (Pahayokea) penningtonorum* Petuch, 1994, length 54 mm; H = *Akleistostoma diegelae* (Petuch, 1994), length 29 mm; I = *Terebraspira seminole* Petuch, 1994, length 130 mm; J = *Siphocypraea (Pahayokea) basingerensis* Petuch, 2004, length 56 mm; K = *Cerithidea briani* Petuch, 1994, length 17 mm.

Siphocypraea (Pahayokea) gabrielleae (Figure 4.28F)
Siphocypraea (Pahayokea) penningtonorum (Figure 4.29G)
Siphocypraea (Pahayokea) rucksorum (Figure 4.28H)
Terebraspira maryae (Figure 4.28E)
Terebraspira seminole (Figure 4.29I) (G)
Pyruella carraheri
Hystrivasum palmerae (Figure 4.29A)
Ventrilia kissimmeensis

During Unit 2 time, the northern end of the Kissimmee Embayment was connected to the Atlantic Ocean by a series of narrow cuts across the St. Lucie Peninsula. This allowed cooler water conditions to enter parts of the embayment, producing marine conditions more similar to the Carolinas than Florida. This cooler marine climate also allowed a large part of the Chowan River Formation (Edenhouse Member) molluscan fauna to invade the Okeechobean Sea. These Virginia and Carolina species remained in the northern end of the Kissimmee Embayment and moved no farther south. Some of the more abundant Chowan River species include:

Gastropoda
Crepidula (Bostrycapulus) chamnessi
Trossulasalpinx moniliferus
Busycon auroraensis (Figure 4.28A)
Busycotypus concinnum (Figure 4.28G)
Brachysycon canaliferum (Figure 4.28J)
Sinistrofulgur adversarium (Figure 4.28B)
Volutifusus auroraensis
Volutifusus typus (Figure 4.28I)
Contraconus petiti (Figure 4.28D)
Bivalvia
Costaglycymeris hummi
Noetia carolinensis
Carolinapecten bertiensis (Figure 4.28C)
Macoma carolinensis

The marine communities of the Kissimmee Embayment during Fruitville Member time were described by the senior author (Petuch, 2004: 190–199) and include the *Pyrazisinus kissimmeensis* Community (mangrove forests and mud flats), the *Siphocypraea kissimmeensis* Community (sand flats and Turtle Grass beds), and the *Siphocypraea penningtonorum* Community (Turtle Grass beds during Unit 2 time). In light of new discoveries in the Kissimmee Valley (at Rucks Pit) over the past two years, the age of some of the mollusks and the structure of some of the communities may need to be revisited.

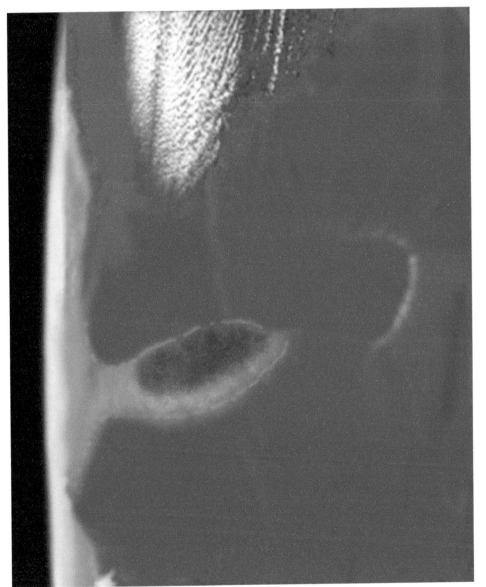

FIGURE 1.3 Dade Subsea, Okeechobean Sea.

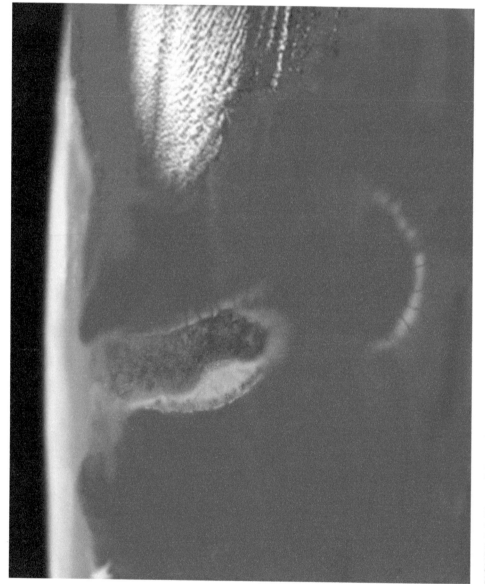

FIGURE 1.4 Tampa Subsea, Okeechobean Sea.

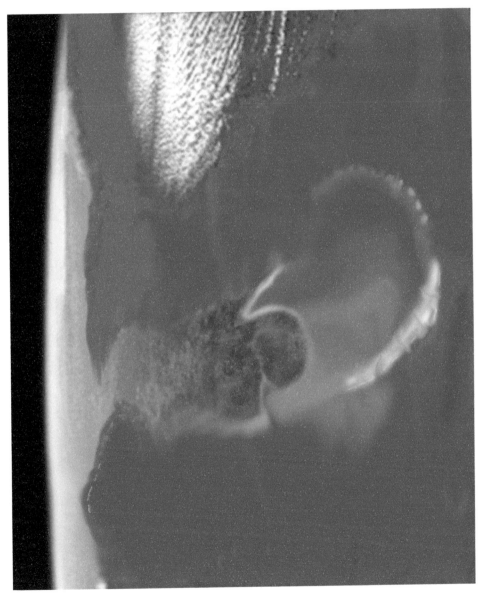

FIGURE 1.5 Arcadia Subsea, Okeechobean Sea.

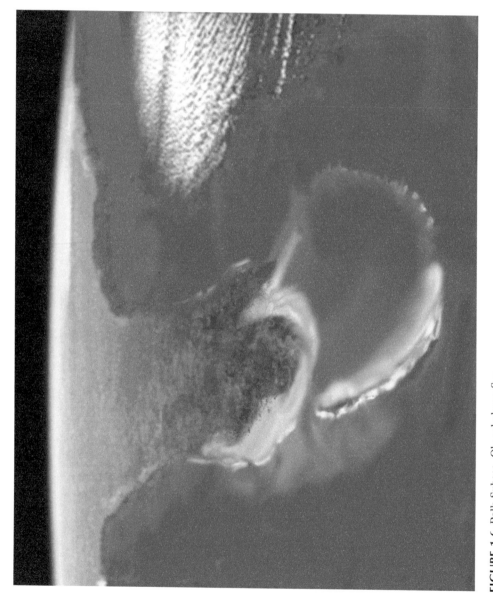

FIGURE 1.6 Polk Subsea, Okeechobean Sea.

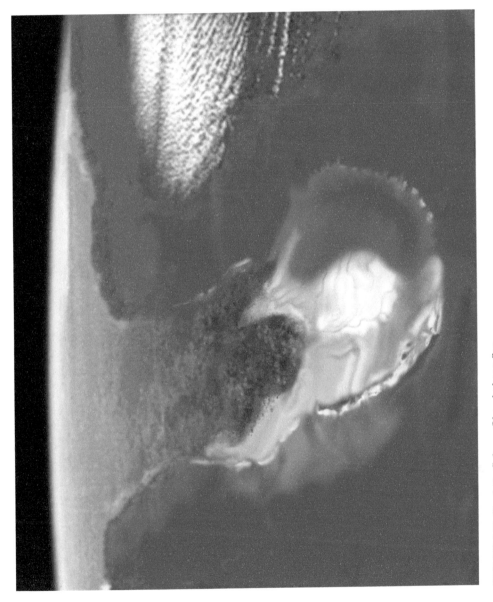

FIGURE 1.7 Charlotte Subsea, Okeechobean Sea.

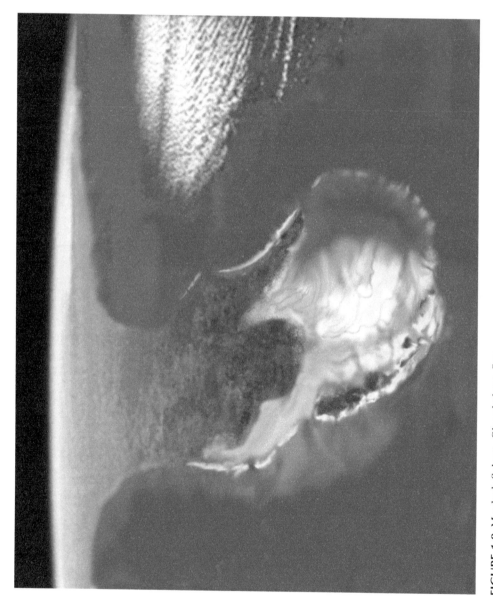

FIGURE 1.8 Murdock Subsea, Okeechobean Sea.

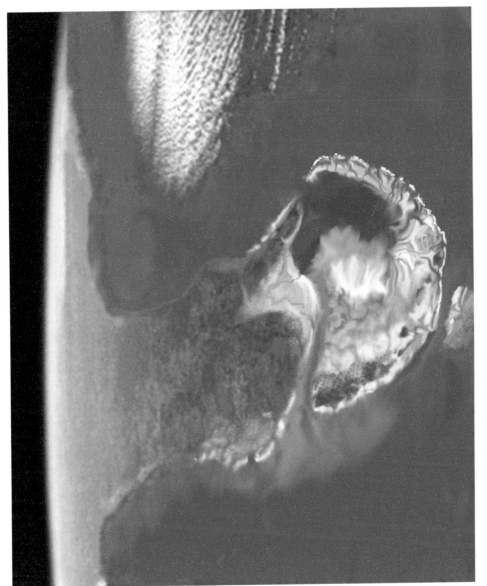

FIGURE 1.9 Tamiami Subsea, Okeechobean Sea.

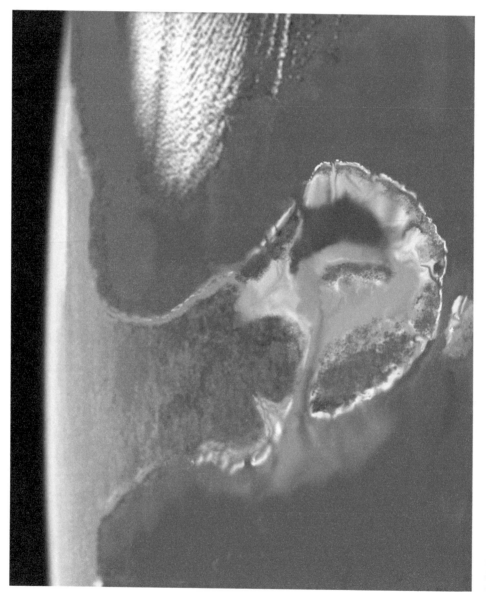

FIGURE 1.10 Caloosahatchee Subsea, Okeechobean Sea.

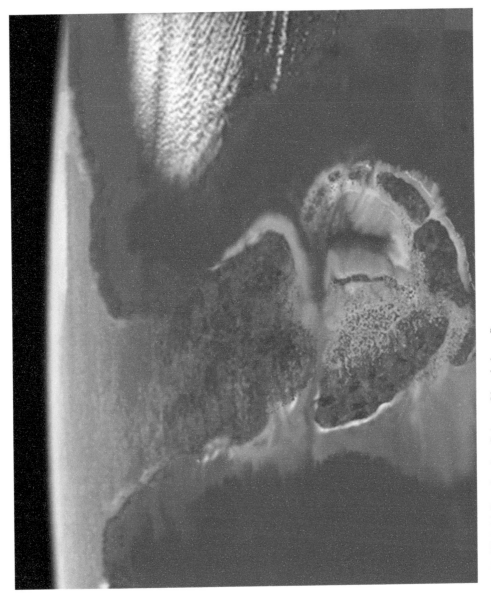

FIGURE 1.11 Loxahatchee Subsea, Okeechobean Sea.

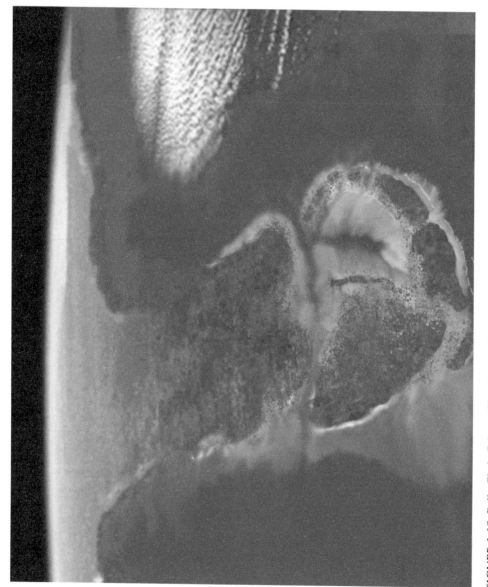

FIGURE 1.12 Belle Glade Subsea, Okeechobean Sea.

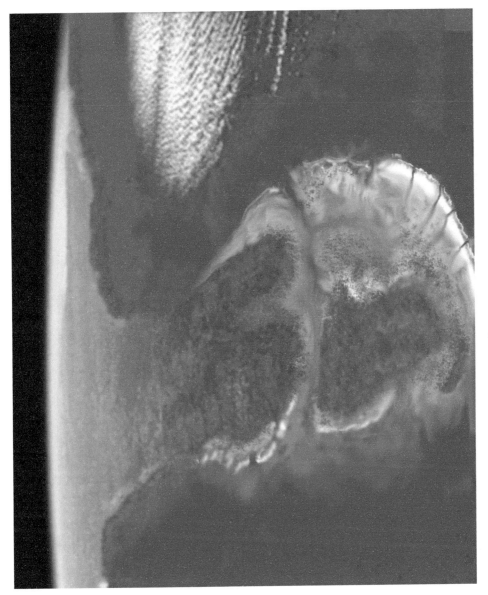

FIGURE 1.13 Lake Worth Subsea, Okeechobean Sea.

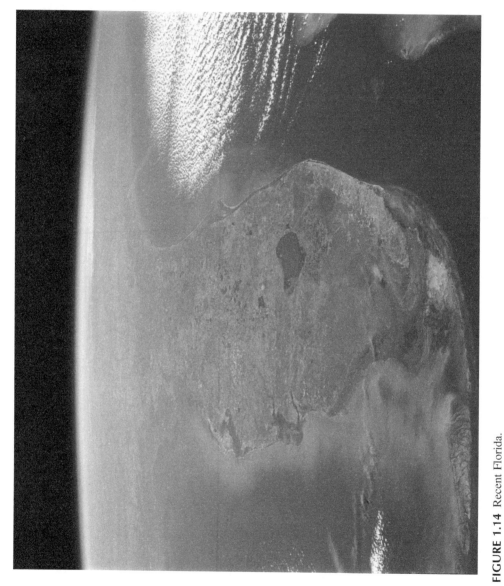

FIGURE 1.14 Recent Florida.

5 Latest Pliocene to Earliest Pleistocene Southern Florida

The last million years of the Pliocene was a time of rapid sea level rises and falls and gradual climatic cooling (Figure 4.2, Chapter 4). Along the eastern coast of the U.S. at this time, many widespread tropical Kissimmean mollusks, such as *Contraconus* and *Pterorhytis*, began to die out, probably in response to the beginning of colder water conditions (Petuch, 1995). This regional extinction is best seen by a comparison of the faunas of the latest Pliocene Chowan River Formation and the earliest Pleistocene James City Formation (Virginia and North Carolina), where these and many other tropical groups disappear at the boundary between the formations. At around 2.4 million years B.P., sea levels plummeted to the lowest they had been since the Zanclean (Figure 4.2), and the Tamiami Subsea ceased to exist.

For over 200,000 years, the Okeechobean Sea basin was emergent and probably housed a series of large freshwater lakes within the Everglades depression. At the very end of the Pliocene, around 2.2 million years B.P., sea levels rose abruptly, and the Okeechobean basin was reflooded with a new marine world, the Caloosahatchee Subsea (named for the Caloosahatchee Formation; see Petuch, 2004) (Figure 5.1). By the Plio–Pleistocene boundary time, the Okeechobean Sea had become essentially landlocked, resembling a saltwater lake surrounded by low islands covered by mangroves and pine forests. Whereas the areas farther north along the Atlantic Coastal Plain had cooler water conditions, the Caloosahatchee Subsea remained warm and tropical due to solar heating of the shallow enclosed basin. Because of this localized, isolated tropical marine environment, the Caloosahatchee Subsea acted as a refugium for many of the regionally extinct Yorktown- and Duplin-type molluscan genera. Whereas groups such as *Contraconus* died out in the northern areas, they flourished within the Caloosahatchee Subsea, producing new endemic species radiations.

PALEOGEOGRAPHY OF THE CALOOSAHATCHEE SUBSEA, OKEECHOBEAN SEA

By the beginning of Caloosahatchee time, the pseudoatoll reef tracts had become covered with sand and rubble and had become large island chains (Figure 5.1). In the west, **Immokalee Island** had enlarged to cover roughly one third of the width of the **Hendry Platform**. Likewise, the narrow Miami and Palm Beach Reef Tracts of Tamiami time had now widened and supported a long chain of islands, the **Miami Archipelago**. The **Cape Sable Bank** was also enlarging at this time and had almost fused to the southern part of Immokalee Island. A new feature, **Miccosukee Island** (named for the Miccosukee Tribe; see Petuch, 2004) had now developed along the eastern edge of the Hendry Platform. This elongated island was covered with mangrove forests and was bounded by coral reefs along its eastern side, adjacent to the deep **Loxahatchee Trough**. As in previous subseas, the pseudoatoll feature was separated from the Florida mainland by the narrow **Caloosahatchee Strait** in the west and the **Loxahatchee Strait** in the east.

The Florida mainland adjacent to the Caloosahatchee Subsea had also become geomorphologically complex during the Plio–Pleistocene boundary time. On the east coast, the most prominent feature was the **Nashua Lagoon System** (named for the Nashua Formation; see Petuch, 2004), a series of narrow, muddy coastal lagoons which formed behind sand-barrier islands (extending northward to the Georgia border). The cuts across the St. Lucie Peninsula, which allowed Chowan

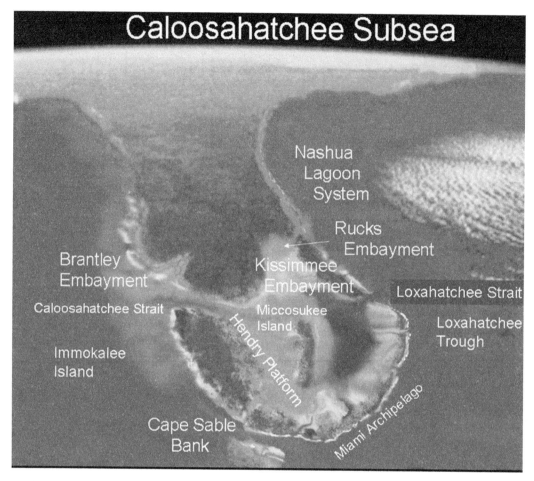

FIGURE 5.1 Simulated space shuttle image (altitude: 100 mi) of the Floridian Peninsula during the early Calabrian Pleistocene, showing the possible appearance of the Caloosahatchee Subsea of the Okeechobean Sea and some of its principal geomorphologic features.

River mollusks to migrate into the Kissimmee Embayment during Tamiami time, had now enlarged into a wide, bay-like feature that connected to the Nashua Lagoon System. This feature, the **Rucks Embayment** (named for the Rucks Pit in Fort Drum, Okeechobee County), was the only area inside the Okeechobean Sea to contain both muddy coastal lagoon conditions and a northern James City Formation-type molluscan fauna.

On the west coast of the Floridian Peninsula, the Myakka Lagoon System was now filling rapidly, being replaced by peaty, organic-rich mud flats, cypress forests, and freshwater marshlands. To the south, in the present-day Peace River Valley, a new broad, bay-like estuarine feature had formed, the **Brantley Embayment** (named for the Brantley Pit, Arcadia, DeSoto County; see Petuch, 2004). This feature, which extended to the east of present-day Arcadia, contained extensive quartz sand flats, Turtle Grass beds, and deep tidal channels filled with coral bioherms. The Okeechobean Sea basin, southern Kissimmee Embayment, and the Brantley Embayment were the main centers of deposition for the Caloosahatchee Formation and its four members, the Fordville, Fort Denaud, Bee Branch, and Ayers Landing. The Rucks Embayment and the Nashua Lagoon System were the centers of deposition for the contemporaneous Nashua Formation and its two members, the Fort Drum and the Rucks Pit. Both formations straddle the Pliocene–Pleistocene boundary.

THE CALOOSAHATCHEE FORMATION, OKEECHOBEE GROUP

HISTORY OF DISCOVERY

The beds exposed at the low-water line along the Caloosahatchee River near Lake Okeechobee have the distinction of being the first geologic unit described from the Everglades area. First noticed and written about by Angelo Heilprin (1886), they were later given the name Caloosahatchee beds by W.H. Dall (1887 and 1892: 140–149). Later, in 1909, Matson and Clapp named them the Caloosahatchee Marl and also described and named the contemporaneous Nashua Marl. It was not until 1958 that the formation was formally described by Jules DuBar. This large, comprehensive work was DuBar's doctoral dissertation, and it contained the first detailed lithologic and paleontologic descriptions, the descriptions of new members, and a review of the depositional paleoenvironments. DuBar followed the convention of the time and referred to the unit as the Caloosahatchee Marl. We here refer to this unit as the Caloosahatchee Formation.

LITHOLOGIC DESCRIPTION AND AREAL EXTENT

From DuBar (1958: 34–35):

> Typically, beds of the Caloosahatchee formation consist of marls composed primarily of quartz sand, silt, and shells. Most of the strata are soft or only slightly indurated, but some are calcareous and very hard, so as closely to approach the nature of true limestone. Most layers are moderately to abundantly fossiliferous, although some, especially sands, are almost or completely barren. Fresh exposures are generally light colored, with white, light gray, cream, and buff predominating. In the subsurface, many sand layers are light green to olive-green. Weathered marls are usually medium to dark gray.

DuBar's "marls" are actually arenaceous or argillaceous calcarenites and calcilutites. These are often interbedded with quartz sand units, frequently packed with molluscan and coral bioclasts. At the Brantley Pit near Arcadia, all four members of the Caloosahatchee Formation were exposed. Lithologies within these units varied from blue-gray sandy lime mud, to light yellow-tan quartz sand with mollusk shells, to white quartz sand mixed with dolosilts. No limestones were present at the quarry, although sandy limestones were reported at quarries 20 km to the south. Other diagnostic lithologies present in the Caloosahatchee include sandy lime muds (arenaceous calcilutites) with large, dense, gray sandy concretions, and peaty sands, often with freshwater snails and clams (naiads), and carbonized wood.

The areal extent of the Caloosahatchee Formation varies with member and reflects sea level fluctuations during Plio–Pleistocene boundary time. The Fordville (Figure 5.2), Fort Denaud (Figure 5.4), and Bee Branch (Figure 5.8) Members all have, essentially, the same areal distributions, with the greatest being during Fort Denaud time. At its greatest development then, the Caloosahatchee deposition extended from southern Pinellas County to Monroe County and from southern Highlands and Okeechobee Counties to Dade County (Figure 5.4). The formation is absent under parts of the Immokalee Rise and Atlantic Coastal Ridge, being represented by undifferentiated sands. During Caloosahatchee time, these areas were islands with terrestrial conditions.

The Caloosahatchee Formation varies in thickness from 3 m at Fort Denaud, Hendry County (DuBar, 1974: Figure 4), to 10 m at the Brantley Pit near Arcadia, DeSoto County, to 15 m at the 10-Mile Bend on the Miami Canal, Palm Beach County, to over 60 m in the Loxahatchee Trough area west of Delray Beach, Palm Beach County (Puri and Vernon, 1964). The surface contours of the Caloosahatchee conform to the bathymetric contours of the Caloosahatchee Subsea. At the Palm Beach Aggregates quarries, in the Loxahatchee Trough area, the formation lies 15–18 m below surface. Along the Miami Canal, bordering the eastern edge of Miccosukee Island, the formation lies only 3 m below surface. At some areas, such as Fort Denaud on the Caloosahatchee River, the formation is less than 2 m below surface and is exposed at the lowest water line during periods of drought.

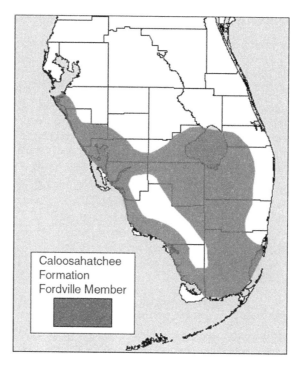

FIGURE 5.2 Areal distribution of the Fordville Member of the Caloosahatchee Formation.

On the sea level curves shown on Figure 4.2 (Chapter 4) and on Figure 6.1 (Chapter 6), the four Caloosahatchee members can be seen to correspond to individual marine transgressions. These incursions were separated from each other by dry, regressive intervals of 100,000–200,000 years, producing obvious erosional unconformities between the individual members. The final reflooding of the Caloosahatchee Subsea during Ayers Landing time coincided with a period of low sea levels (Figure 6.1, Chapter 6). Because of this, the last Caloosahatchee deposition was much more restricted in distribution (Figure 5.10), covering only about 70% of the area of the underlying Bee Branch Member.

STRATOTYPE

Until now, there has been no true stratotype for the Caloosahatchee Formation, only broad areas that contain exposures along the low water line of the Caloosahatchee River. Two researchers' observations: " ... The typical exposures of the formation are on the Caloosahatchee River at and near LaBelle, Hendry County ..." (Cooke, 1945: 215) and " ... The type exposures of the Caloosahatchee marl are found along the Caloosahatchee River between Fort Denaud and Ortona Locks ..." (DuBar, 1958: 35). We here designate the stratotype to be the section described by DuBar (1958: 236–237) at his station A36 (NW ¼ SW ¼, Sec. 11, T. 45 S., R. 28 E., Hendry County, Florida) on the " ... left bank of the Caloosahatchee River about 0.9 mile upstream (east) from the bridge at Fort Denaud." The Caloosahatchee stratotype section includes DuBar's units 1–7 (with 8 being the Fort Thompson Formation and 9 and 10 being the Pamlico Formation). This section is still available for study, and crops out at 1 m below the top of the section that extends to the water line.

AGE AND CORRELATIVE UNITS

Based on the presence of key widespread index fossils such as the gastropods *Ximeniconus waccamawensis*, *Ventrilia betsiae*, *Ventrilia elizabethae*, *Sinum multiplicatus*, *Ilyanassa wilmingtonensis*, and *Ilyanassa granifera* (see Petuch, 1994) and the bivalves *Mercenaria permagna*,

Macrocallista greeni, and *Dinocardium hazeli*, the Caloosahatchee Formation is now known to span the time of the latest Piacenzian Pliocene through the entire Calabrian Pleistocene. The Fordville Member is latest Piacenzian, the Fort Denaud Member straddles the Plio–Pleistocene boundary, and the Bee Branch and Ayers Landing Members are both Calabrian. The formation is correlative with the Colerain Beach Member of the Chowan River Formation (Fordville Member only) and the James City Formation of Virginia and northern North Carolina, the Bear Bluff and Waccamaw Formations of southern North Carolina and northern South Carolina, and the Fort Drum and Rucks Pit Members of the Nashua Formation of eastern Florida.

CALOOSAHATCHEE INDEX FOSSILS

As mentioned previously in this chapter, the enclosed warm Caloosahatchee Subsea acted as a refugium for many of the Kissimmeean-aged molluscan genera. Those that survived the extinction event at the end of Tamiami time underwent new species radiations, producing sets of highly endemic faunas. Some of the Tamiami survivors that speciated during the times of the four Caloosahatchee members include the gastropod genera and subgenera *Pyrazisinus, Okeechobea, Pahayokea, Siphocypraea, Macrostrombus, Echinofulgur, Terebraspira, Turbinella*, and *Contraconus*. Successive species from each of these genera, together, give the fauna of each member its own distinctive appearance. These assemblages can be used to determine the boundaries between members.

THE FORDVILLE MEMBER, CALOOSAHATCHEE FORMATION

History of Discovery

The senior author, in describing the stratigraphy at the Ashland Petroleum and Asphalt Corporation (APAC) pit in Sarasota (Petuch, 1982), first noticed that the uppermost fossiliferous bed (Unit 1) contained a molluscan fauna that was different from those contained in the layers below. A more detailed analysis showed that these fossils belonged to the Caloosahatchee Formation and that the sediments ("beach deposit") were also quite different from those of the underlying Tamiami Formation. The locality of this measured section (Figure 4.3, Chapter 4) was on the eastern side of the APAC pit, and this Caloosahatchee layer was thin (1 m) and poorly developed. Subsequently, William Lyons encountered the same unit on the western side of the APAC pit, but there it was much better developed, being over 4 m thick and containing nine distinct beds (Lyons, 1992: 133–159).

Although containing numerous classic Caloosahatchee index fossils such as *Turbo rhectogrammicus, Cerithioclava caloosaensis, Ximeniconus waccamawensis*, and *Cancellaria conradiana*, Unit 1 was also found to contain a large number of very distinct restricted species. Some of these were misidentified by Lyons, who gave these endemic forms the names of living Recent species (Lyons, 1992: Figure 1 to Figure 80). This newly discovered fauna has also been found elsewhere around the Everglades area, demonstrating that Unit 1 has a wider distribution than originally thought.

Based on the distinctive and complex lithology, the wide areal extent, and the chronologically restricted molluscan fauna, we here informally name this lowermost set of Caloosahatchee beds the Fordville Member. This new member takes its name from Fordville, a small agricultural hamlet (Mennonite farms) 3 km southeast of the APAC pit on Fruitville Road. The Quality Aggregates quarries, on the other side of I-75 from the APAC pit (east), are adjacent to Fordville and also contain exposures of the new member. Because this new member is present at all the quarries in the Fordville area, and because the names Fruitville and Sarasota have already been used for other older units, we chose the village as the only viable possibility for an informal geologic name.

Lithologic Description and Areal Extent

As shown by Lyons (1992: 135), the lithology of the Fordville Member is quite variable. Typically, the member is composed of quartz sands with varying amounts of peats, carbonized wood, and particulate organic matter. Some quartz sand and peat layers are packed with well-preserved mollusks, often the bivalves *Carditamera*, *Anomalocardia*, and *Mulinia*. Large calcareous biohermal masses of sabellariid colonial polychaete worms are often present, forming discontinuous reef-like structures. The quartz sands are variably colored, ranging from white, to brown, to black. Some sand units contain minor amounts of clays, usually sepiolite and illite. The mollusks and sabellariid "worm rock" masses are colored brown or blackish-brown, often weathering to a light reddish-brown. Fordville beds in the southern areas of its distribution contain a higher carbonate fraction, usually variable amounts of calcilutites and dolosilts. These beds typically contain large molluscan bioclasts and are colored a light gray or blue-green. Limestones are absent in this member.

As can be seen on Figure 5.2, the Fordville Member covers a smaller area than do the subsequent Fort Denaud and Bee Branch Members. Over its range, the Fordville varies in thickness from 4 m at the stratotype at Sarasota, to 2 m in the Brantley Pit, Arcadia, DeSoto County, to 5 m along the Miami Canal (canal dredging), Palm Beach County. In the Fordville area of Sarasota, the member is only 1–3 m below surface whereas at the Brantley pit, it occurs at depths of 15 m below surface. Along the Miami Canal, the member occurs at 15–20 m below surface and at Mule Pen Quarry at East Naples, Collier County, it lies 5–8 m below surface.

Stratotype

We here designate Unit 1 on the western side of the APAC pit (see Figure 4.3, Chapter 4) as the stratotype of the Fordville Member. This site, located on Richarson Road, adjacent to I-75, is now flooded as a lake in a Sarasota County public park. A map of the stratotype (T. 36 S., R. 18 E., Sec. 12, Sarasota County) was given by Lyons (1992: Figure 1). This locality is approximately 3 km northwest of Fordville, Sarasota County (on Fruitville Road). The Fordville stratotype beds are 2–3 m below the mean lake surface and can be accessed by scuba diving.

Age and Correlative Units

Based on the presence of key widespread index fossils, such as the gastropods *Pterorhytis conradi*, *Littorinopsis lindae*, and *Terebraspira cronleyensis* and the bivalve *Conradostrea sculpturata*, the Fordville Member is now known to date from the latest Piacenzian Pliocene. The new member is correlative with the Colerain Beach Member of the Chowan River Formation of Virginia and northern North Carolina, the Bear Bluff Formation of southern North Carolina and northern South Carolina, and the basal beds of the Fort Drum Member of the Nashua Formation of eastern Florida.

Fordville Index Fossils

This previously overlooked member contains a distinctive suite of chronologically restricted species that can be used to demarcate the member boundaries. Besides a large component of endemic species, the Fordville fauna also contains a small but noteworthy component of upper Chowan River Formation species. Some of the more abundant and indicative Fordville index taxa include:

Endemic Floridian species
Gastropoda
Trivia permagna (Figure 5.3G)
Jenneria lindae (carbonate-rich facies only)
Siphocypraea dimasi (Figure 5.3I) (carbonate-rich facies only)
Siphocypraea mulepenensis (Figure 5.3F) (carbonate-rich facies only)
Hexaplex trippae (Figure 5.3C)
Pterorhytis seminola (Figure 5.3A)
Terebraspira calusa (Figure 5.3D) (listed as *T. scalarina* by Lyons, 1992: Figure 47)
Melongena unnamed species (listed as *M. subcoronata* by Lyons, 1992: Figure 42)

FIGURE 5.3 Index fossils for the Fordville Member of the Caloosahatchee Formation. A = *Pterorhytis seminola* Olsson and Petit, 1964, length 27 mm; B = *Oliva (Strephona) briani* Petuch, 1994, length 86 mm; C = *Hexaplex trippae* (Petuch, 1991), length 81 mm; D = *Terebraspira calusa* Petuch, 1994, length 97 mm; E = *Hystrivasum martinshugari* Petuch, 1994, length 92 mm; F = *Siphocypraea mulepenensis* Petuch, 1991, length 70 mm; G = *Trivia (Niveria) permagna* Johnson, 1910, length 22 mm; H = *Littorinopsis lindae* Petuch, 1994, length 15 mm; I = *Siphocypraea dimasi* Petuch, 1998, length 42 mm; J = *Pterorhytis conradi* (Dall, 1890), length 59 mm; K = *Terebraspira cronleyensis* (Gardner, 1948), length 138 mm.

Hystrivasum horridum subspecies
Hystrivasum martinshugari (Figure 5.3)
Oliva (Strephona) briani (Figure 5.3B)
Contraconus tryoni subspecies

Species shared with the Colerain Beach Member of the Chowan River Formation
Littorinopsis lindae (Figure 5.3H) (new name for *L. lineata* Emmons, 1858, not *L. lineata*
    Gmelin, 1791; listed as *L.* species by Lyons, 1992: Figure 10)
Pterorhytis conradi (Figure 5.3J)
Terebraspira cronleyensis (Figure 5.3K)
Ximeniconus waccamawensis
Bivalvia
Conradostrea sculpturata

THE FORT DENAUD MEMBER, CALOOSAHATCHEE FORMATION

History of Discovery

In his 1958 study of the Caloosahatchee, DuBar recognized three distinct sections: an "upper shell bed," a middle Bee Branch Member (limestone), and a "lower shell bed" (DuBar, 1958: 50–64, Figure 23). He later (1958a; 1974) designated the upper bed the Ayers Landing Member and the lower bed the Fort Denaud Member. Of the three members, the Fort Denaud is the most stratigraphically complex and contains the richest molluscan fauna of the entire formation. This member contains the Caloosahatchee mollusk species described and illustrated by Heilprin (1886).

DuBar (1958; 1974) was the first person to describe the paleoecology and depositional environments of the Caloosahatchee Formation. He noted that, of the three members, the Fort Denaud contained the most types of environments, including coral patch reefs (DuBar, 1974: 217), oyster bioherms (the "basal oyster biostrome"), mud flats containing *Cyrtopleura* and *Panopea* deep-burrowing bivalves (the "*Cyrtopleura costata* Faunizone"), *Vermicularia* worm gastropod bioherms, brackish water calcarenite beds with *Rangia* clams, and freshwater environments.

Lithologic Description and Areal Extent

From DuBar (1958: 51), for his "lower beds," later named the Fort Denaud Member:

> Typically, the lower beds are light colored, with cream, white, and light gray predominating, although some units are mottled yellow-brown by limonite staining, and all are some shade of gray on weathered surfaces. At almost all localities, the beds are comprised of sandy (quartz) and silty marls. Locally, the percentages of sand-sized particles is high enough to warrant describing the rock as calcareous sandstone … Calcareous concretions of various sizes and shapes are abundant. Many are small fossiliferous nodules that occur in random manner scattered in the bed, whereas some are joined so as to form an arborescent network extending throughout an entire bed … Many of the beds are fossiliferous. Almost every facies and all units carry an abundant, varied, and well preserved fauna. Mollusks are dominant, but foraminifers are well represented, and ostracodes, barnacles, and bryozoans are common.

As discussed earlier in the chapter, the Fort Denaud has the widest areal extent of all the Caloosa-hatchee members (Figure 5.4), extending from Pinellas, southern Highlands, southern Okeechobee, and southern Martin Counties in the north to Monroe and Dade Counties in the south. Throughout its range, the Fort Denaud is also the thickest member, being over 10 m thick at St. Petersburg, Pinellas County, 4 m thick along the Miami Canal in Palm Beach County, 30 m thick west of Delray Beach, Palm Beach County, and 1–3 m thick along the central area of the Caloosahatchee River.

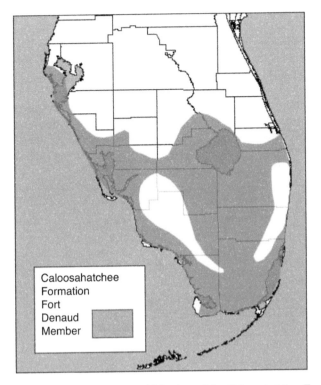

FIGURE 5.4 Areal distribution of the Fort Denaud Member of the Caloosahatchee Formation.

Stratotype

From DuBar (1958: 50–51):

> … Exposures of the lower beds (Fort Denaud Member) are most typically developed along a stretch of the river (Caloosahatchee) extending between points 0.5 and 3.5 mi upstream from Fort Denaud in Hendry County. Here, the beds rise higher above the water level than elsewhere, are usually thicker, and are laterally more extensive.

We here select one of DuBar's Fort Denaud area collecting localities (A39) as the stratotype (SW ¼, NW ¼, Sec. 10, T. 43 S., R. 28 E., Hendry County, Florida): " … right bank of the Caloosahatchee River about 325 yards downstream from the bridge at Fort Denaud and 170 yd downstream from Jack's Branch." Here, the Fort Denaud occurs at 3 m below the top of the section.

Age and Correlative Units

Based on the presence of widespread key index fossils, such as the gastropods *Ximeniconus waccamawensis*, *Ilyanassa granifera*, *Ilyanassa wilmingtonensis*, *Globinassa schizopyga*, *Sinum multilineatum*, and *Triplofusus acmaensis*, and the bivalves *Argopecten vicenarius* and *Carolinapecten solaroides*, the Fort Denaud Member is now known to span the Pliocene–Pleistocene boundary, at approximately 2 million years B.P. (shown along with the sea level curves on Figure 4.2, Chapter 4 and Figure 6.1, Chapter 6). The member is correlative with the lower beds of the James City Formation of Virginia and northern North Carolina, the lower beds of the Waccamaw Formation of southern North Carolina and northern South Carolina, and the upper beds of the Fort Drum Member of the Nashua Formation of eastern Florida.

Fort Denaud Index Fossils

DuBar (1974: Table 5) gave a listing of characteristic macrofossils for his Caloosahatchee members, showing that many species were present in all three. Subsequent studies by the senior author (Petuch, 1991; 1994) have shown that many of DuBar's chronologically widespread taxa actually represented successions of chronologically restricted descendant species that evolved rapidly from their Fort Denaud ancestors. Although containing a small component of widespread Waccamaw and James City species, the majority of the Fort Denaud molluscan fauna is composed of Floridian endemics, restricted to the Caloosahatchee Subsea. The coral fauna, similarly, contains a large percentage of endemic species (Petuch, 2004: 220). Some of the more abundant and indicative index taxa (most illustrated in Petuch, 1994) include:

Gastropoda
Turbo (Taenioturbo) rhectogrammicus (Figure 5.6D)
Astraea (Lithopoma) precursor
Tegula calusa
Apicula apicalis
Bactrospira perattenuata
Modulus caloosahatcheensis (Figure 5.7K)
Littorinopsis caloosahatcheensis (Figure 5.7J)
Littorinopsis seminole
Cerithidea jenniferae (Figure 5.7E)
Cerithidea xenos (Figure 5.7L)
Pyrazisinus scalatus (Figure 5.6F)
Pyrazisinus ecarinatus
Cerithioclava caloosaense
Siphocypraea problematica (Figure 5.5A)
Siphocypraea (Okeechobea) philemoni (Figure 5.6K)
Siphocypraea (Pahayokea) josiai (Figure 5.7I)
Jenneria richardsi (Figure 5.7F)
Trivia incerta
Malea springi (Figure 5.5H)
Macrostrombus leidyi (Figure 5.5D)
Strombus keatonorum (Figure 5.6E)
Chicoreus calusa
Chicoreus susanae
Phyllonotus evergladesensis (Figure 5.6B)
Pterorhytis wilsoni (Figure 5.5E) (also found in the Waccamaw Formation)
Subpterynotus cf. *textilis* (Figure 5.7A) (The real *S. textilis* is a Miocene fossil from the Baitoa Formation of the Dominican Republic.)
Eupleura intermedia
Eupleura calusa
Fasciolaria (Cinctura) apicina
Fasciolaria calusa
Fasciolaria monocingulata
Heilprinia caloosaensis
Liochlamys bulbosa (Figure 5.5G)
Terebraspira scalarina (Figure 5.5J)
Echinofulgur echinatum (Figure 5.5F)
Melongena caloosahatcheensis
Busycoarctum rapum

FIGURE 5.5 Index fossils for the Fort Denaud Member of the Caloosahatchee Formation. A= *Siphocypraea problematica* Heilprin, 1886, length 62 mm; B = *Calusaconus spuroides* (Olsson and Harbison, 1953), length 37 mm; C = *Contraconus tryoni* (Heilprin, 1886), length 158 mm; D = *Macrostrombus leidyi* (Heilprin, 1886), length 147 mm; E = *Pterorhytis wilsoni* Petuch, 1994, length 40 mm; F = *Echinofulgur echinatum* (Dall, 1890), length 88 mm; G = *Liochlamys bulbosa* (Heilprin, 1886), length 95 mm; H = *Malea springi* Petuch, 1989, length 186 mm; I = *Hystrivasum horridum* (Heilprin, 1886), length 99 mm; J = *Terebraspira scalarina* (Heilprin, 1886), length 160 mm.

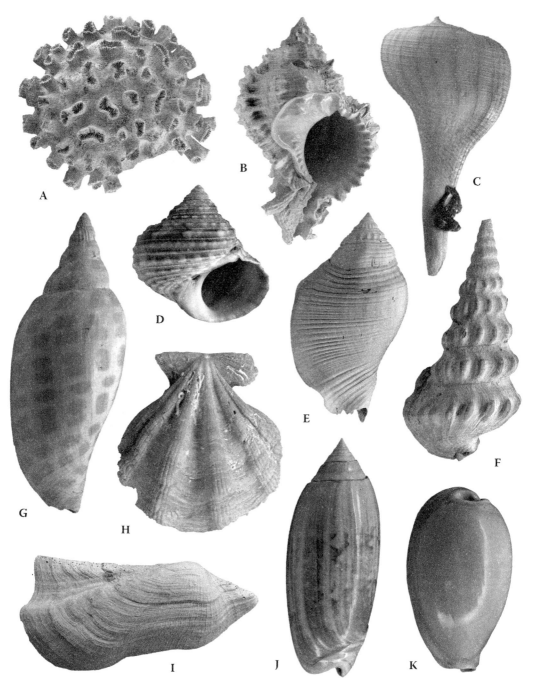

FIGURE 5.6 Index fossils for the Fort Denaud Member of the Caloosahatchee Formation. A = *Dichocoenia eminens* Weisbord, 1974, length 146 mm; B = *Phyllonotus evergladesensis* Petuch, 1994, length 86 mm; C = *Sinistrofulgur caloosahatcheensis* Petuch, 1994, length 108 mm; D = *Turbo (Taenioturbo) rhectogrammicus* Dall, 1892, length 43 mm; E = *Strombus keatonorum* Petuch, 1994, length 73 mm; F = *Pyrazisinus scalatus* (Heilprin, 1886), length 78 mm; G = *Scaphella floridana* Heilprin, 1886, length 140; H = *Stralopecten caloosaensis* (Dall, 1898), length 65 mm; I = *Arcoptera wagneriana* (Dall, 1898), length 98 mm; J = *Oliva (Strephona) erici* Petuch, 1994, length 87 mm; K = *Siphocypraea (Okeechobea) philemoni* Fehse, 1997, length 61 mm.

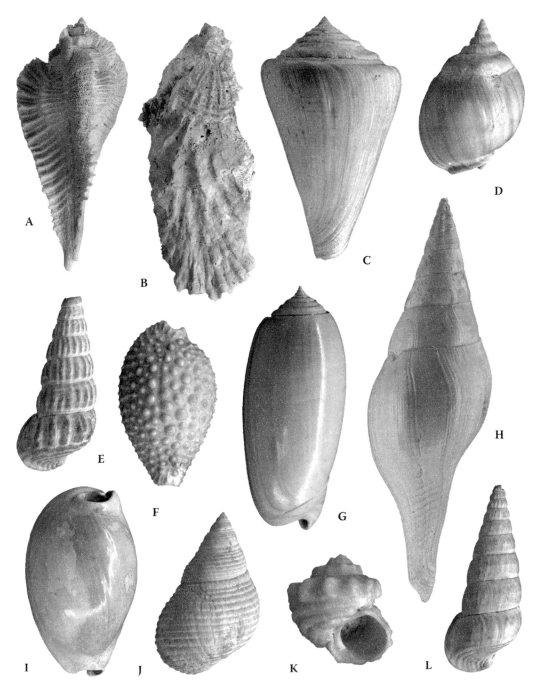

FIGURE 5.7 Index fossils for the Fort Denaud Member of the Caloosahatchee Formation. A = *Subpterynotus* cf. *textilis* (Gabb, 1873), length 66 mm; B = *Crassostrea labellensis* (Olsson and Harbison, 1953), length 119 mm; C = *Contraconus osceolai* Petuch, 1991, length 88 mm; D = *Globinassa roseae* (Petuch, 1991), length 23 mm; E = *Cerithidea jenniferae* Petuch, 1994, length 27 mm; F = *Jenneria richardsi* Olsson, 1967, length 28 mm; G = *Oliva (Porphyria) paraporphyria* Petuch, 1991, length 67 mm; H = *Turbinella regina* Heilprin, 1886, length 286 mm; I = *Siphocypraea (Pahayokea) josiai* Fehse, 1997, length 43 mm; J = *Littorinopsis caloosahatcheensis* (Petuch, 1991), length 18 mm; K = *Modulus caloosahatcheensis* Petuch, 1994, length 10 mm; J = *Cerithidea xenos* Petuch, 1991, length 25 mm.

Sinistrofulgur caloosahatcheensis (Figure 5.6C)
Sinistrofulgur palmbeachensis
Pyruella planulatum
Pyruella soror
Fulguropsis floridanus
Globinassa floridana
Globinassa roseae (Figure 5.7D)
Globinassa schizopyga (also found in the Waccamaw Formation)
Ilyanassa granifera (also found in the James City and Waccamaw Formations)
Ilyanassa palmbeachensis
Ilyanassa wilmingtonenesis (also found in the Waccamaw Formation)
Scalanassa evergladesensis
Scalanassa olssoni
Turbinella regina (Figure 5.7H)
Turbinella scolymoides
Hystrivasum horridum (Figure 5.5I)
Scaphella floridana (Figure 5.6G)
Cancellomorum macgintyi
Morum floridanum
Cancellaria conradiana
Perplicaria perplexa
Pleioptygma lineolata
Oliva (Porphyria) paraporphyria (Figure 5.7G)
Oliva (Strephona) erici (Figure 5.6J)
Oliva (Strephona) roseae
Calusaconus spuroides (Figure 5.5B)
Contraconus osceolai (Figure 5.7C)
Contraconus tryoni (Figure 5.5C)
Gradiconus parkeri
Jaspidiconus wilsoni
Seminoleconus diegelae (also found in the James City Formation)
Ximeniconus waccamawensis (also found in the Waccamaw and James City Formations)
Hindsiclava perspirata
Knefastia lindae
Bivalvia
Arcoptera wagneriana (Figure 5.6I) (also found in the Bee Branch Member)
Leptopecten irremotis
Lindapecten harrisi
Stralopecten caloosaensis (Figure 5.6H)
Crassostrea labellensis (Figure 5.7B) (also found in the Bee Branch Member)
Cnidaria–Scleractinia
Dichocoenia caloosahatcheensis
Dichocoenia eminens (Figure 5.6A)
Thysanus floridanus

The marine communities of the Fort Denaud Member were described by the senior author (Petuch, 2004: 212–222) and include the *Pyrazisinus scalatus* Community (mangrove forests and intertidal mud flats), the *Siphocypraea problematica* Community (Turtle Grass beds), the *Arcoptera wagneriana* Community (shallow sand-bottom lagoons), and the *Dichocoenia eminens* Community (coral reefs and coral bioherms).

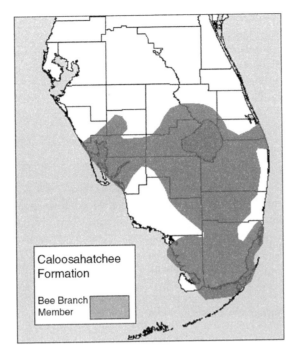

FIGURE 5.8 Areal distribution of the Bee Branch Member of the Caloosahatchee Formation.

THE BEE BRANCH MEMBER, CALOOSAHATCHEE FORMATION

History of Discovery

During DuBar's stratigraphic survey of the Caloosahatchee River, he recognized a distinct, prominent limestone layer midway in the section. This resistant bed frequently formed a wide, thick ledge that jutted out from the section just above the low-water line. The best-developed exposures were found to be in the vicinity of the confluence of a small tributary of the Caloosahatchee River, the Bee Branch, and the member name derives from this stream. Petuch (1994) recognized Bee Branch sediments and fossils from the Miami Canal dredgings, Palm Beach County, and later found large exposures in the Brantley Pit near Arcadia, DeSoto County.

Lithologic Description and Areal Extent

From DuBar (1958: 58, condensed here):

> The name Bee Branch member is here proposed to designate a relatively hard, solution-riddled marine limestone or marl unit of the Caloosahatchee marl. It is typically exposed along the Caloosahatchee River in Hendry County near the confluence with the tributary called Bee Branch ... Two principal facies of the Bee Branch member and many minor variations of each have been recognized. One facies is represented by the Bee Branch deposits of the type locality which form a massive, hard calcareous bed ... It is riddled with large and small solution holes that are filled with matrix from overlying beds. Commonly, small pockets of softer marly material occur within the bed ... Most of the weathered surfaces are dark gray or black, but fresh surfaces are buff or cream colored. Yellow-brown mottling by limonite stain is common ... The other dominant facies differs from the type locality in being softer, less consolidated, more arenaceous (quartz sand), and more distinctly concretionary. Generally, it does not form prominent ledges. The concretions contained in it are hard, dense calcareous nodules which locally occur in isolated manner but more commonly form an arborescent network extending throughout the entire bed.

As discussed earlier, the Bee Branch has the second-largest areal distribution of the four Caloosahatchee members (Figure 5.8). It ranges from southernmost Sarasota, western DeSoto, southeastern Highlands, southern Okeechobee, and southern Martin Counties in the north, to Monroe and Dade Counties in the south. Throughout its range, the Bee Branch Member is relatively thin, being 2–3 m thick in the Brantley Pit, near Arcadia, DeSoto County, 0.5–2 m thick along the Caloosahatchee River near LaBelle, Hendry County, to 3 m thick along the Miami Canal in western Palm Beach County.

Stratotype

DuBar (1958: 59) gave the type locality as the bank of the Caloosahatchee River in a stretch between his stations A22 and A26 in Hendry County (… left bank of the Caloosahatchee River about 0.8 mile downstream from the bridge at LaBelle" and 1025 yd downstream from the "sharp bend in the canal" (near the confluence with Bee Branch). DuBar (1974: 217) later stated that his station A70 is the stratotype of the Bee Branch but gives no details or references. We here designate, as the lectostratotype of the Bee Branch Member, DuBar's (1958) station A22 (NW $^{1}/_{4}$ Sec. 5, T. 43 S., R. 29 E., Hendry County, Florida) at the "left bank Caloosahatchee River about 0.8 mile downstream from the bridge at LaBelle." The member is exposed at 2 m below the top of the section.

Age and Correlative Units

Based on its stratigraphic position within the Caloosahatchee Formation, the Bee Branch is here dated as being of early Calabrian Pleistocene age. The member is correlative with the upper beds of the James City Formation of Virginia and northern North Carolina, the upper beds of the Waccamaw Formation of southern North Carolina and northern South Carolina, and the lower beds of the Rucks Pit Member of the Nashua Formation of eastern Florida.

Bee Branch Index Fossils

Like the Fort Denaud Member, the Bee Branch contains a highly endemic fauna that is restricted to the Everglades area. Because much of the Bee Branch is composed of limestone, the following list is of index fossils that are primarily associated with the unconsolidated calcarenite facies. The richest Bee Branch faunas encountered to date were found along the Miami Canal, Palm Beach County, in excavations west of the Okeelanta sugar mill, and at the Brantley Pit near Arcadia, DeSoto County. Some of the more abundant and indicative Bee Branch species include:

Gastropoda
Siphocypraea problematica subspecies
Siphocypraea (Okeechobea) brantleyi (Figure 5.9F)
Siphocypraea (Pahayokea) aspenae (Figure 5.9A)
Macrostrombus brachior (Figure 5.9I)
Chicoreus sarae
Phyllonotus labelleensis (Figure 5.H)
Fasciolaria seminole
Fasciolaria (Cinctura) lindae (Figure 5.9G)
Terebraspira labelleensis (Figure 5.9B)
Echinofulgur palmbeachensis (Figure 5.9D)
Sinistrofulgur labelleensis
Fulguropsis elongatus
Hystrivasum horridum subspecies
Turbinella wendyae
Cancellomorum obrienae

FIGURE 5.9 Index fossils for the Bee Branch Member of the Caloosahatchee Formation. A = *Siphocypraea (Pahayokea) aspenae* Petuch, 2004, length 61 mm; B = *Terebraspira labelleensis* Petuch, 1994, length 101 mm; C = *Dauciconus gravesae* (Petuch, 1994), length 29 mm; D = *Echinofulgur palmbeachensis* Petuch, 1994, length 77 mm; E = *Cancellaria clewistonensis* Olsson and Harbison, 1953, length 51 mm; F = *Siphocypraea (Okeechobea) brantleyi* Petuch, 2004, length 56 mm; G = *Fasciolaria (Cinctura) lindae* Petuch, 1994, length 42 mm; H = *Phyllonotus labelleensis* Petuch, 1994, length 86 mm; I = *Macrostrombus brachior* (Petuch, 1994), length 138 mm.

Cancellaria clewistonensis (Figure 5.9E)
Contraconus species (similar to *C. osceolai*)
Dauciconus gravesae (Figure 5.9C)
Gradiconus ronaldsmithi

The marine communities of the Bee Branch Member were described by the senior author (Petuch, 2004: 222–224) and include: the *Anomalocardia caloosana* Community (shallow sand-bottom lagoons) and the *Siphocypraea brantleyi* Community (Turtle Grass beds).

THE AYERS LANDING MEMBER, CALOOSAHATCHEE FORMATION

History of Discovery

In his stratigraphic survey of the Caloosahatchee River area, DuBar (1958) recognized an "upper shell bed" within the Caloosahatchee sequence. He later (1958a; 1974) named this the Ayers Landing Member, after an old boat landing near the Fort Denaud bridge. Along the river, this uppermost set of beds can be seen to lie directly on the limestone ledges of the Bee Branch Member. In the late 1980s, the senior author recognized Ayers Landing sediments and index fossils in excavations at the Griffin Brothers Pit adjacent to the Holey Land Wildlife Management Area in southwesternmost Palm Beach County. Many of these were later described (Petuch, 1991; 1994), and these have subsequently been collected at numerous localities, such as along the Miami Canal and at the Brantley Pit.

Lithologic Description and Areal Extent

From DuBar (1958: 62–63, condensed here):

> At many localities the bed is very arenaceous and the calcium carbonate content is mainly in the form of mollusk shells. Thick beds are generally only slightly consolidated and are easily eroded. The sands (quartz) are usually tan to yellow-brown, but the fossils are nearly white. At some localities, the unit is better consolidated, much more calcareous, concretionary, and finer grained. The color is usually light buff, but weathers to dark gray … . At all localities, the member is fossiliferous. The faunal remains consist mostly of mollusk shells, but corals, foraminifers, ostracodes, barnacles, and bryozoans are common … thin beds occur near the top of the section and appear to be of fresh water origin; other beds clearly represent a brackish water environment.

At the stratotype, the member is composed of three distinct beds: an upper tan-colored, slightly concretionary, unconsolidated marine calcarenite with sparse marine mollusks, a middle sandy, tan, freshwater limestone with freshwater snails, and a lower sandy, cream-colored marine calcarenite (weathering to gray) filled with fossil mollusks.

As can be seen on Figure 5.10, the Ayers Landing Member has the smallest areal distribution of the four Caloosahatchee members. It extends from southern Sarasota, southwestern DeSoto, Charlotte, Glades, and western Palm Beach Counties in the north, to Dade County in the south. From the sea level curves shown on Figure 6.1 (Chapter 6), the Ayers Landing can be seen to have been deposited during the smallest marine transgression of the Pleistocene, and this explains its limited distribution. Over this small range, the member varies in thickness from 1 m at the Brantley Pit near Arcadia, DeSoto County, to greater than 3 m along the Caloosahatchee River in Hendry County, to 2 m at the Griffin Brothers Pit in southwesternmost Palm Beach County, to 1 m in canal digs west of the Miami Canal, west of the Okeelanta sugar mill, Palm Beach County.

Stratotype

From DuBar (1958: 62, condensed here): " … The best exposures of the shell bed occur along the Caloosahatchee River in Hendry County between stations A26 and A35, and this is considered to be the type locality." The stratotype is here recognized as DuBar's station A35, in the Ayers Landing

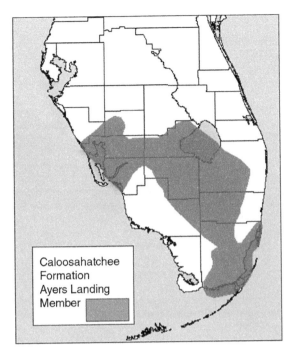

FIGURE 5.10 Areal distribution of the Ayers Landing Member of the Caloosahatchee Formation.

area, "90 yards upstream from Turtle Branch" (SW ¼, NW ¼, Sec. 11, T. 43 S., R. 28 E., Hendry County, Florida). DuBar commented that at this locality " ... there occurs an excellent exposure of the upper Caloosahatchee shell bed that contains, in addition to a profusion of mollusks and coral heads, a good vertebrate fauna including the teeth of *Equus (Equus)* sp. cf. *E. (E.) leidyi.*"

Age and Correlative Units

Based on the stratigraphic position within the Caloosahatchee sequence, the Ayers Landing Member dates from the late Calabrian Pleistocene. The member is correlative with the uppermost beds of the James City Formation of Virginia and northern North Carolina, the uppermost beds of the Waccamaw Formation of southern North Carolina and northern South Carolina, and the uppermost beds of the Rucks Pit Member of the Nashua Formation of eastern Florida.

Ayers Landing Index Fossils

This youngest Caloosahatchee member (approximately 1.7 million years) contains the last assemblages of the Kissimmean Mollusk Age faunas. Some of the classic Kissimmean genera that appear for the last time within the Ayers Landing Member include *Contraconus*, *Siphocypraea*, *Hystrivasum*, and *Echinofulgur*. Their disappearance at the end of Ayers Landing time can be used to demarcate the boundary between the Caloosahatchee Formation and the younger Bermont Formation (see Chapter 6). Like the Bee Branch Member, the Ayers Landing contains a highly endemic fauna that is restricted to the Everglades area. (Most of these are illustrated and described in Petuch, 1994.) Some of the more abundant and indicative species include:

Gastropoda
Pyrazisinus intermedius
Siphocypraea griffini (Figure 5.11G)
Macrostrombus jonesorum (Figure 5.11B)

FIGURE 5.11 Index fossils for the Ayers Landing Member of the Caloosahatchee Formation. A = *Ximeniconus robertsi* (Olsson and Harbison, 1953), length 38 mm; B = *Macrostrombus jonesorum* (Petuch, 1994), length 194 mm; C = *Echinofulgur griffini* Petuch, 1991, length 105 mm; D = *Hystrivasum griffini* Petuch, 1994, length 39 mm; E = *Scaphella oleiniki* Petuch, 1994, length 63 mm; F = *Oliva (Strephona) jenniferae* Petuch, 1994, length 48 mm; G = *Siphocypraea griffini* Petuch, 1991, length 39 mm; H = *Turbinella lindae* Petuch, 1994, length 178 mm; I = *Contraconus heilprini* Petuch, 1994, length 32 mm; J = *Contraconus scotti* Petuch, 1994, length 59 mm.

Phyllonotus martinshugari
Liochlamys griffini
Melongena chickee
Echinofulgur griffini (Figure 5.11C)
Pyruella eismonti
Pyruella ovoidea
Fulguropsis griffini
Scaphella griffini
Scaphella oleiniki (Figure 5.11E)
Scaphella tomscotti
Turbinella lindae (Figure 5.11H)
Hystrivasum griffini (Figure 5.11D)
Oliva (Strephona) jenniferae (Figure 5.11F)
Cancellaria calusa
Contraconus heilprini (Figure 5.11I)
Contraconus mitchellorum
Contraconus scotti (Figure 5.11J)
Ximeniconus robertsi (Figure 5.11A)

The marine communities of the Ayers Landing Member were described by the senior author (Petuch, 2004: 224–226) and include the *Oliva jenniferae* Community (shallow sand-bottom lagoons) and the *Siphocypraea griffini* Community (Turtle Grass beds).

THE NASHUA FORMATION, OKEECHOBEE GROUP

History of Discovery

The pioneer Floridian geologists Matson and Clapp first reported on a fossiliferous "white shell marl" (calcarenite) from the vicinity of Nashua, Putnam County, on the eastern side of the St. John's River south of Palatka (Matson and Clapp, 1909). Further analyses showed that this site contained some Caloosahatchee-aged fossils, and they assigned a Pliocene age to the beds within the Nashua area. Because these beds contained more clay and fewer carbonate fines than did the Caloosahatchee "marl," and because the molluscan fossils of the St. John's River beds were of a cooler water, Waccamaw-type, Matson and Clapp formally named these beds the Nashua Marl (later Nashua Formation). Some later workers, such as Cooke and Mossom (1929: 152), considered these differences insignificant and included the Nashua in the Caloosahatchee Formation. In light of new information from better sections, we here consider the Nashua to be a separate formation, contemporaneous with the Caloosahatchee Formation from farther south within the Everglades Basin.

In 1940, Sydney Stubbs of the Florida Geological Survey reported on a Nashua assemblage of fossils (as "Caloosahatchee") from a 30 m-deep well at Sanford, Seminole County. Stubbs also described similar fossils and sediment types from wells in DeLeon Springs and Daytona Beach, Volusia County, and these were cited by Cooke (1945: 227, from unpublished memoranda). More recently, the senior author (Petuch, 1994) reported on a Nashua fauna from Rucks Pit in Fort Drum, northern Okeechobee County and described several new key index fossils. A molluscan fauna and an accompanying set of lithofacies, similar to those at Fort Drum, were also found at the Dickerson Corporation, Inc., Indrio quarry, west of Fort Pierce, St. Lucie County. All of these localities demonstrate that the Nashua Formation has a much larger areal distribution than was originally thought.

Lithology and Areal Extent

The Nashua Formation was deposited within a narrow coastal lagoon (the Nashua Lagoon System) that had formed behind a chain of sand barrier islands, which extended from Georgia to St. Lucie

County. The southern end of the Nashua Lagoon, in the vicinity of present-day Indian River and St. Lucie Counties, was connected to the northern end of the Kissimmee Embayment of the Caloosahatchee Subsea by a series of narrow channels. These channels effectively cut the St. Lucie Peninsula into a chain of wide islands. The northernmost of these channels emptied into a large, shallow bay at the northern end of the Kissimmee Embayment (the Rucks Embayment). This was the center of deposition for the Everglades component of Nashua sedimentation.

As pointed out by Cooke (1945: 214), the molluscan fossils of the Nashua Formation are cooler water species, and many of them are found in the contemporaneous Waccamaw Formation of the Carolinas. This indicates that, during Caloosahatchee Subsea time, the open Atlantic was much cooler than the interior of the enclosed Okeechobean Sea. The presence of cooler water species, mixed with some tropical forms, demonstrates that the depositional environments of the Nashua were oceanographically intermediate between those of the high tropical Caloosahatchee and temperate Waccamaw and James City Formations. These cool water conditions are reflected in the complete absence of hermatypic coral reefs and bioherms and by the preponderance of terriginous siliciclastics. The connecting channels into the Rucks Embayment allowed Waccamaw-type water conditions, sediments, and ecosystems to exist at the northern end of the Kissimmee Embayment … but no farther south.

Typically, the Nashua Formation comprises sets of interbedded units with varying lithologies, including unconsolidated sandy (quartz) calcarenites and calcilutites with variable amounts of clay minerals (mostly sepiolite and illite), quartz sand with densely packed calcareous bioclasts made up mostly of lagoonal and estuarine bivalves (Figure 5.12), shell hash coquinas with spar cementing, and cavity-filled, very sandy, spar-cemented limestone (a vuggy arenaceous sparite), at some localities grading into a sandstone (Figure 5.12 and Figure 5.15). At the Rucks Pit ("Fort Drum Crystal Mine"), this sandy sparite contains abundant molluscan shell calcite geodes filled with dogtooth spar. The unconsolidated sandy calcarenites (like those at the stratotype) are white or pale gray with white or pale tan molluscan bioclasts. The spar-cemented sandy limestones (or limy sandstones) are typically dark gray with honey-gold-colored calcite crystals forming as vug filling or as molluscan geodes. True coquinas are the least common lithology, and are found in thin (less than 0.5 m thick) beds or lenses and are of a pale tan or cream color. The clay component of the unconsolidated beds increases in abundance in the northern extent of the formation. In the southern areas of Nashua deposition, the dark gray sandy limestones often occur in thick, massive units and are mined commercially for road construction.

The Nashua Formation occurs in a narrow strip that parallels the present coastline, extending inland 30–60 km (Figure 5.13). In the south, the formation extends into the northern end of the Okeechobee Plain. The Nashua ranges from St. Johns, Putnam, Flagler, Volusia, Seminole, easternmost Orange, easternmost Osceola and Brevard Counties south to Indian River, northernmost Okeechobee, St. Lucie, and Martin Counties. Throughout its range, the Nashua is variable in thickness, being 5 m thick at the stratotype in Putnam County, 3.5 m thick at Sanford, Seminole County, 8–10 m thick at the Rucks Pit, Fort Drum, Okeechobee County, and 10–12 m thick in the Indrio Pit on Indrio Road west of Fort Pierce, St. Lucie County. Along the coast, the Nashua comes to within 1.5–5 m from land surface. Farther inland (to the west), the Nashua can occur at depths ranging from 20–30 m (such as at Sanford).

The Nashua Formation contains two lithostratigraphic members: a lower Fort Drum Member (latest Piacenzian Pliocene and the Plio–Pleistocene boundary time) and an upper Rucks Pit Member (Calabrian Pleistocene). These new members are informally described in the following sections.

STRATOTYPE

The type locality of Matson and Clapp's Nashua Formation is an exposure "one-quarter mile" south of Nashua, Putnam County, near the St. John's River. Here, 5 km of a "white shell marl" was exposed and was overlain unconformably by 1.5 m of white sand. Because this locality is now covered and lost, we here designate a locality adjacent to Nashua, that was described by Cooke

FIGURE 5.12 View of the north side of the Rucks Pit, Fort Drum, Okeechobee County, showing the three main beds of the Fort Drum Member of the Nashua Formation. At the top is a thick layer of massive gray sandy limestone (or limy sandstone) filled with *Mercenaria* clam calcite geodes. Below the gray sandy limestone is a thick unconsolidated bed of single *Mercenaria permagna* valves. Below the clam layer, to the water surface, is a thick layer of unconsolidated shell hash containing well-preserved large molluscan fossils such as *Carolinapecten senescens*, *Brachysycon willcoxi*, and *Scaphella brennmortoni*. Photo by Anton Oleinik.

and Mossom (1929: 225) " … in a gully … just south of the center of Sec. 28, T. 11 S., R. 26 E. (Putnam County, Florida), in the bluff east of the St. John's River." The stratotype exposure probably represents the upper beds of the Nashua, correlative with the Rucks Pit Member in southern Florida.

AGE AND CORRELATIVE UNITS

Based on the presence of widespread key index fossils such as the gastropods *Brachysycon willcoxi*, *Brachysycon amoenum*, *Scaphella brennmortoni*, and *Ximeniconus waccamawensis*, and the bivalves *Carolinapecten senescens*, *Noetia limula*, and *Conradostrea lawrencei*, the Nashua Formation is now known to span the time frame of the latest Piacenzian Pliocene, the Plio–Pleistocene boundary, and the entire Calabrian Pleistocene. The formation is exactly correlative with the Colerain Beach Member of the Chowan River Formation of Virginia and northern North Carolina, the Bear Bluff and Waccamaw Formations of southern North Carolina and northern South Carolina, and the Caloosahatchee Formation of southern Florida.

THE FORT DRUM MEMBER, NASHUA FORMATION

History of Discovery

In 1992, Dr. Thomas Scott, the Assistant State Geologist, discovered calcite spar geodes within a gray sandy limestone in the Rucks Pit at Fort Drum, Okeechobee County (also known as the

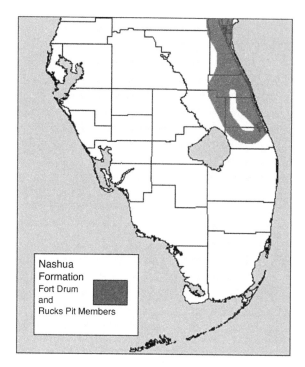

FIGURE 5.13 Areal distribution of the Nashua Formation and its two members, the Fort Drum and the Rucks Pit.

Dickerson Corporation, Inc., Fort Drum Mine, or the Fort Drum Crystal Mine). Besides the presence of calcite geodes, the mine was of special interest to geologists in that it was continuously pumped, allowing for *in situ* collecting of fossil specimens and for direct study of the exposed geological formations. The Rucks Pit calcite geodes were highly unusual, because they had formed as aggregations of large dogtooth spar growing inside fossilized bivalves, mostly pairs of *Mercenaria* (Figure 5.14B and Figure 5.14C), and large gastropods, mostly *Busycon* (Figure 5.14D). Nothing like this had ever been found before within the Everglades region.

Dr. Scott also noted the presence of thick beds of fossils interbedded in a white calcarenite and observed that they were especially well preserved and contained a large number of species not normally found within the Everglades area. The senior author was brought to the mine by Dr. Scott in 1993 and soon discovered that the molluscan fauna contained not only new endemic species but also species originally described from the Waccamaw, James City, and Caloosahatchee Formations. These well-preserved mollusks were contained within the white, unconsolidated upper layer within the quarry (Figure 5.15), and this unit is undoubtedly the "white shell marl" of Matson and Clapp. The lower dark gray sandy limestone, however, had never before been reported or described. In late 2004, the senior author informally gave names to these two Nashua members in a pamphlet published as a field guide for visitors to the Fort Drum Crystal Mine. The proposed lower member, the Fort Drum, is here informally described.

Lithologic Description and Areal Extent

The Fort Drum Member consists of interbedded units of variable lithologies, including thick layers of a dark gray, very sandy (quartz) limestone or a dark gray, limy sandstone cemented by calcite microspar (arenaceous sparites), dense beds of the bivalve *Mercenaria* or busyconid gastropods in a matrix of gray or brown unconsolidated quartz sand (both seen in Figure 5.12), dark brown muddy sand filled with small broken shell bioclasts ("shell hash"), or lenses of light tan or cream-colored coquina (small shell fragments similar to the "shell hash" layers). Typically, the dark gray sandy limestone is extremely vuggy and moldic, often containing the partially dissolved remnants of *Mulinia* and *Noetia* valves. At certain localities, such as at Fort Drum, the main gray sandy limestone bed contains large molluscan bioclasts, primarily recalcified pairs of the bivalve *Mercenaria permagna*. These closed pairs are filled with large, honey-tan or golden-yellow crystals of iron-stained calcite (Figure 5.14B and Figure 5.14C).

At the stratotype area in the Rucks Pit, the massive gray sandy limestone bed rests on a thick bed of large, perfectly preserved, closely packed single valves of *Mercenaria permagna* (apparently winnowed by tidal currents; Figure 5.12). This distinctive unconsolidated clam bed grades into the gray sandy limestone above and into the muddy "shell hash" layer below. The same massive gray arenaceous sparite-limy sandstone of the Fort Drum Member is also exposed within the Indrio Pit (Dickerson Corporation, Inc.) west of Fort Pierce, but there it lacks the *Mercenaria* geodes. In the Indrio Pit, the unconsolidated beds below the gray sandy limestone are not filled with *Mercenaria* clams but, instead, are replaced by beds of beautifully preserved busyconid whelks (several genera). North of Indian River County, the Fort Drum Member becomes predominantly a coarse quartz sand unit, dark brown or gray in color, with variable amounts of molluscan bioclasts.

The areal extent of the Fort Drum Member is the same as that of the entire formation (Figure 5.13). In the area covered by this book, the member is present only in northeastern Okeechobee, St. Lucie, and northern Martin Counties. At the Rucks Pit, the Fort Drum Member averaged 6 m in thickness, whereas at the Indrio Pit it was over 7 m thick.

Stratotype

We here designate the stratotype of the Fort Drum Member as the lower beds on the southern wall of the Rucks Pit (Figure 5.15) on Cemetery Road (NE 304th Street), 3 km east of U.S. Highway 441, and the town of Fort Drum, Okeechobee County, Florida. The type section is 6.2 m below land surface. The new member is named for the town of Fort Drum, Okeechobee County.

Age and Correlative Units

Based on the presence of widespread key index fossils, such as the gastropods *Brachysycon willcoxi* and *Ximeniconus waccamawensis* and the bivalves *Carolinapecten senescens* and *Mercenaria permagna*, the Fort Drum Member is now known to be of latest Piacenzian Pliocene to early Calabrian Pleistocene age. The member is correlative with the Colerain Beach Member of the Chowan River Formation of Virginia and northern North Carolina, the Bear Bluff Formation and lower beds of the Waccamaw Formation of southern North Carolina and northern South Carolina, and the Fordville and Fort Denaud Members of the Caloosahatchee Formation of southern Florida.

Fort Drum Index Fossils

Besides *Mercenaria* valves and busyconid whelks, very few other types of mollusks are known from the Fort Drum Member. This low diversity of fauna is probably the result of the harsh oceanographic conditions within the estuaries of the Nashua Lagoon System. These would have included cold winter water temperatures and fluctuating salinities. Most of the Fort Drum mollusks are wide-ranging species, being found from northern North Carolina to southern Florida. Some of the more commonly encountered and indicative Fort Drum species include (W, also found in the Waccamaw Formation):

Gastropoda
Eupleura calusa (also found in the Caloosahatchee Formation)
Brachysycon willcoxi (Figure 5.14H) (W)

Busycon duerri (Figure 5.14D and Figure 5.14G) (also found in the Caloosahatchee Formation. This species was originally thought to have come from the younger Ayers Landing
Member of the Caloosahatchee Formation (Petuch, 2004: 225), but is now known to be
an older species, from the Fort Denaud Member and equivalents. The name for the whelk
in the Ayers Landing communities should be changed to *Busycon rucksorum*, as specimens
of that species have recently been collected in Ayers Landing beds in Palm Beach County).
Sinistrofulgur pamlico (Figure 5.14F) (W)
Scaphella brennmortoni (Figure 5.14I) (W)
Ximeniconus waccamawensis (Figure 5.14J) (W) (also found in the Caloosahatchee Formation)

Bivalvia

Carolinapecten senescens (Figure 5.14E) (W)
Mercenaria permagna (Figure 5.14A) (W) (also found in the Caloosahatchee Formation)

A single marine community was described from the Fort Drum Member by the senior author
(Petuch, 2004: 208-212), the *Mercenaria permagna* Community (intertidal sand flat and sand bar
environment).

THE RUCKS PIT MEMBER, NASHUA FORMATION

History of Discovery

The excavations at Fort Drum in the early 1990s uncovered large exposures of a white, sandy
unconsolidated lime mud (calcilutite) filled with small *Mulinia* bivalve shells (Figure 5.15). The
molluscan assemblage was much richer and more diverse than that of the underlying gray sandy
limestones and clam beds. Like the Fort Drum Member mollusks, most of the shells of the upper
white layer were found to be of northern origin, primarily shared with the James City and Waccamaw
Formations. The lithology of this unit and its molluscan fauna closely resemble the Nashua type
section in Putnam County and may be correlative.

Lithologic Description and Areal Extent

The Rucks Pit Member typically is composed of various mixtures of quartz sand and lime mud
packed with mollusk shells (arenaceous bioclastic calcilutite). The quartz sand and fine particulate
carbonates are white, cream-white, or light gray, weathering to white in exposed sections. In some
facies, quartz sand dominates and contains a minor percentage of clay minerals, mostly sepiolite
and montmorillonite. The percentages of clay minerals increase in the northern part of the member,
whereas the carbonate fines dominate in the southern part. In other facies, the small estuarine
bivalve *Mulinia lateralis* forms densely packed beds with only a small amount of calcilutites and
quartz sand. Intercalated throughout the *Mulinia* beds are thin layers composed of beds of the
scallop *Carolinapecten solaroides*, small bioherms of the oyster *Conradostrea lawrencei*, and larger
bioherms of the ahermatypic coral *Septastrea* species. At the Indrio Pit west of Fort Pierce, the
Rucks Pit Member is represented by a thick (2–3 m), white quartz sand layer with scattered beach-
worn mollusk shells. The basal part of the *Mulinia* beds frequently indurates into a pale gray sandy
coquina but only forming thin, scattered layers.

The areal extent of the Rucks Pit Member is the same as that of the entire formation (Figure 5.13).
In the area covered by this book, the member is present only in northeastern Okeechobee, St. Lucie,
and northern Martin Counties. At Fort Drum, the member averaged 3 m in thickness and at the
Indrio Pit west of Fort Pierce, it was 3–4 m thick. The section given by Cooke (1945: 226) from
the City Marl Pit near DeLand, Volusia County, contained an exposure of the Rucks Pit Member
at 4 m below land surface. The lithology (" ... creamy yellow sandy shell marl ... ") closely
resembles that of the Rucks Pit stratotype.

FIGURE 5.14 Index fossils for the Fort Drum Member of the Nashua Formation. A = *Mercenaria permagna* (Conrad, 1838), length 126 mm; B, C = Half specimen of a *Mercenaria permagna* dogtooth spar calcite geode, showing recalcified exterior and crystal-filled interior, length 90 mm; D = cross-section of a recalcified *Busycon duerri*, showing calcite crystal growth in interior, length 96 mm; E = *Carolinapecten senescens* (Dall, 1898), length 82 mm; F = *Sinistrofulgur pamlico* Petuch, 1994, length 128 mm; G = *Busycon duerri* Petuch, 1994, length 159 mm (differs from the older Fruitville *B. auroraensis* in having finer and less-pronounced spiral sculpture and a higher, stepped spire); H = *Brachysycon willcoxi* (Gardner, 1948), length 158 mm; I = *Scaphella brennmortoni* Olsson and Petit, 1964, length 142 mm; J = *Ximeniconus waccamawensis* (B. Smith, 1930), length 40 mm.

FIGURE 5.15 View of the southern side of the Rucks Pit, Fort Drum, Okeechobee County, showing the two members of the Nashua Formation, the lower (gray sandy limestone) Fort Drum Member and the upper (white sandy calcarenite) Rucks Pit Member. These represent the type sections of the two new members. Above the white Rucks Pit Member (partially covered by vegetation) is a thick bed of the Coffee Mill Hammock Member of the Fort Thompson Formation (sandy mangrove peat facies).

Stratotype

We here designate the stratotype of the Rucks Pit Member as the southern wall of the Rucks Pit (also Dickerson Corporation, Inc., Fort Drum Mine) (Figure 5.15) on Cemetery Road (NE 304th Street), 3 km east of U.S. Highway 441 and the town of Fort Drum, Okeechobee County, Florida. The type section is 4 m below land surface. The new member is named for the Rucks Pit, Fort Drum, Okeechobee County.

Age and Correlative Units

Based on the presence of several widespread key index fossils, such as the gastropods *Brachysycon amoenum*, *Pyruella bladenense*, and *Volutifusus halscotti* and the bivalves *Noetia limula*, *Conradostrea lawrencei*, and *Dinocardium hazeli*, the Rucks Pit Member is now known to span the middle and late Calabrian Pleistocene. The member is correlative with the upper beds of the James City Formation of Virginia and northern North Carolina, the upper beds of the Waccamaw Formation of southern North Carolina and northern South Carolina, and the Bee Branch and Ayers Landing Member of the Caloosahatchee Formation of southern Florida.

Rucks Pit Index Fossils

The Rucks Pit Member molluscan fauna, found within the Everglades area only in the northern Okeechobee Plain, comprises three separate components: an endemic component of restricted Nashua species, a component of northern James City and Waccamaw species, and a small component of tropical species shared with the Caloosahatchee Formation. Some of the more abundant and indicative species include:

Endemic Nashua species
Gastropoda
Fasciolaria (Cinctura) rucksorum (Figure 5.16H)
Busycon rucksorum (Figure 5.16A)
Busycotypus scotti (Figure 5.16E)
Sinistrofulgur yeehaw (Figure 5.16C)
Oliva (Strephona) rucksorum

Species shared with the James City and Waccamaw Formations
Gastropoda
Brachysycon amoenum (Figure 5.16B)
Pyruella bladenense
Triplofusus acmaensis (also found in the Caloosahatchee Formation)
Volutifusus halscotti (Figure 5.16I)
Strioterebrum petiti
Bivalvia
Noetia limula
Conradostrea lawrencei (Figure 5.16F)
Dinocardium hazeli

Species shared with the Caloosahatchee Formation
Gastropoda
Urosalpinx rucksorum (Figure 5.16J)
Oliva (Porphyria) paraporphyria
Oliva (Strephona) roseae
Ventrilia rucksorum (Figure 5.16G)
Bivalvia
Carolinapecten solaroides (Figure 5.16D)

A single marine community was described from the Rucks Pit Member by the senior author (Petuch, 2004: 210), the *Mulinia lateralis* Community (shallow muddy coastal lagoon environment).

FIGURE 5.16 Index fossils for the Rucks Pit Member of the Nashua Formation. A = *Busycon rucksorum* Petuch, 1994, length 168 mm (differs from the older Fort Drum Member *B. duerri* in having much finer, more reduced spiral sculpture and by having a lower, more conical spire; differs from the living *B. carica* in having more pronounced spiral sculpture); B = *Brachysycon amoenum* (Conrad, 1875), length 112 mm; C = *Sinistrofulgur yeehaw* Petuch, 1994, length 160 mm; D = *Carolinapecten solaroides* (Heilprin, 1886), length 175 mm; E = *Busycotypus scotti* Petuch, 1994, length 155 mm; F = *Conradostrea lawrencei* Ward and Blackwelder, 1987, length 80 mm; G = *Ventrilia rucksorum* Petuch, 1994, length 29 mm; H = *Fasciolaria (Cinctura) rucksorum* Petuch, 1994, length 115 mm; I = *Volutifusus halscotti* Petuch, 1994, length 115 mm; J = *Urosalpinx rucksorum* Petuch, 1994, length 21 mm.

6 Middle Pleistocene Southern Florida

The mid-Pleistocene (1.6 million to 400,000 years B.P.) was a time of rapidly changing climates and sea levels. During some periods of time, such as 1 million years B.P., the climate approximated that seen in southern Florida today and sea level was close to that of the present (Figure 6.1). During other times, such as 850,000 years B.P., sea level plummeted to over 150 m below that of the present, and the entire Florida Platform became emergent. At that time, southern Florida probably had a climate similar to southern Georgia. These wildly fluctuating climates and sea levels had a devastating effect on the marine communities of the Everglades Pseudoatoll. This was manifested in the abrupt elimination of many prominent groups and the rapid evolution of the surviving taxa (Petuch, 1995).

The eustatic fluctuations of the middle Pleistocene came in two large pulses, the first starting at about 1.4 million years B.P. and the second around 750,000 years B.P. (Figure 6.1). These correspond to the Aftonian and Yarmouthian Interglacial Stages. Between the two interglacial warm times, a severe glacial time occurred (the Kansan Glacial Stage), and the Everglades area was emergent for 200,000 years. Throughout Kansan times, the Loxahatchee Trough area of the pseudoatoll became a huge freshwater lake (Lake Okeelanta, described later in this chapter) surrounded by semiarid terrestrial conditions. As can be seen in Figure 6.1, the two major marine transgressions were complex, each having at least two secondary pulses separated by brief sea level drops.

The first major transgression during Aftonian times flooded the Everglades Basin with a new marine system, the Loxahatchee Subsea (named for the town of Loxahatchee, Palm Beach County; Petuch, 2004) (Figure 6.2), whereas the second, during Yarmouthian times, produced the Belle Glade Subsea, named for the city of Belle Glade, Palm Beach County (Petuch, 2004) (Figure 6.12). During the time span of these two subseas, biogenic deposition increased dramatically and rapidly filled the central lagoon area of the pseudoatoll. Throughout this time, the narrow Loxahatchee Trough remained the only deepwater area within the shrinking pseudoatoll. This enclosed, land-locked feature acted as a refugium for many tropical marine organisms from Caloosahatchee times. These survivors, mixed together with newly evolved organisms, gave the marine communities of the Loxahatchee and Belle Glade Subseas a strange and unique appearance.

The complex eustatic patterns of the middle Pleistocene manifested themselves in equally complex patterns of deposition within the Everglades area. The units that resulted from these complex patterns are here grouped together as a single large geologic unit, the Bermont Formation. Its lower member, the Holey Land Member, was deposited during the time of the Loxahatchee Subsea. The middle member, the Okeelanta Member, was deposited within the environs of Pleistocene Lake Okeelanta during the Kansan glacial time. The upper member, the Belle Glade Member, was deposited during the time of the Belle Glade Subsea. The multiple lithofacies seen in the members of the Bermont Formation reflect the complex oceanographic conditions of the Loxahatchee and Belle Glade Subseas and the surrounding Atlantic Ocean and Gulf of Mexico. Besides the Arcadia Formation of late Oligocene and early Miocene times (Chapter 2), the Pleistocene Bermont is the only other Everglades formation to have been deposited within two different Okeechobean subseas.

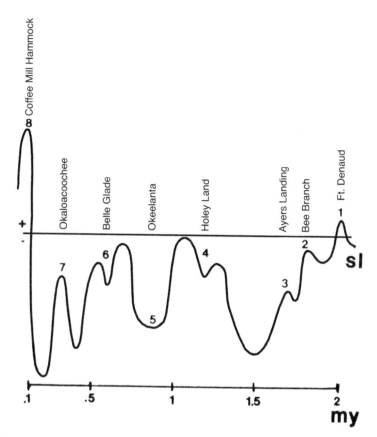

FIGURE 6.1 Sea level curve for the Pleistocene, showing major eustatic fluctuations. Present sea level (sl) is used as a reference standard. Time is in millions of years (my). Numbers correspond to the times of deposition of formations and members, during sea level highs. Formations: (1, 2, 3) = Caloosahatchee Formation; (4, 5, 6) = Bermont Formation; (7, 8) = Fort Thompson Formation. Members: 1 = Fort Denaud Member, Caloosahatchee Formation; 2 = Bee Branch Member, Caloosahatchee Formation; 3 = Ayers Landing Member, Caloosahatchee Formation; 4 = Holey Land Member, Bermont Formation; 5 = Okeelanta Member, Bermont Formation; 6 = Belle Glade Member, Bermont Formation; 7 = Okaloacoochee Member, Fort Thompson Formation; 8 = Coffee Mill Hammock Member, Fort Thompson Formation. (Taken from the Oxygen Isotope Model of Krantz [1991] and from Vail et al. [1977].)

PALEOGEOGRAPHY OF THE LOXAHATCHEE SUBSEA, OKEECHOBEAN SEA

For over 300,000 years, the Everglades area remained above sea level after the retreat of the Caloosahatchee Subsea. During the intervening terrestrial interval, the Everglades contained a large freshwater lake (or possibly a series of smaller lakes), presaging present-day Lake Okeechobee. This early Pleistocene paleolake, Lake Immokalee (named for the city of Immokalee, Collier County; Petuch, 2004: 228) is preserved as a thin (0.5 m) layer of freshwater limestone, filled with freshwater snails, on the surface of the Ayers Landing Member of the Caloosahatchee Formation (see Chapter 5). This paleolake limestone has been found at several localities, including Mule Pen Quarry, East Naples, Collier County, and the Holey Land Levee on the Miami Canal, Palm Beach County (observed *in situ* during water drawdowns).

At around 1.4 million years B.P., the Loxahatchee Subsea began to flood Lake Immokalee and the Everglades Basin. This new subsea enlarged the geomorphologic features that had developed

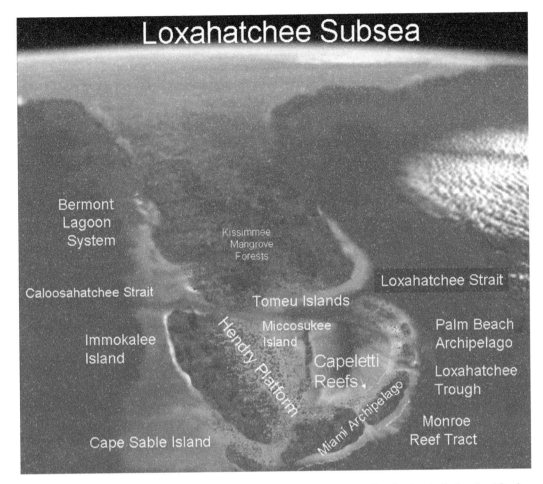

FIGURE 6.2 Simulated space shuttle image (altitude: 100 mi) of the Floridian Peninsula during the Aftonian Pleistocene, showing the possible appearance of the Loxahatchee Subsea of the Okeechobean Sea and some of its principal geomorphologic features.

within the predecessor Caloosahatchee Subsea. As can be seen in Figure 6.2, **Immokalee Island** had expanded to cover most of the **Hendry Platform**, leaving only a small strip of mangrove islands between it and **Miccosukee Island**. Similarly, the Kissimmee Embayment had filled with mangrove jungles (the **Kissimmee Mangrove Forests**), and its northern half was completely emergent. The Cape Sable Bank had now become a large island (**Cape Sable Island**) and was separated from Immokalee Island by a shallow area filled with small mangrove islands.

The eastern side of the pseudoatoll was also changing rapidly during the time of the Loxahatchee Subsea. Here, the **Miami Archipelago** and **Palm Beach Archipelago** were now large, wide chains of islands covered with forests of pine trees and palmettos. On the western side of these islands, within the quiet, solar-heated waters of the **Loxahatchee Trough**, a series of large coral bioherms were forming (the **Capeletti Reefs**, named for the Capeletti Brothers Quarries in Miami). Similarly, a long line of zonated coral reefs, the **Monroe Reef Tract** (named for Monroe County), was forming along the southern edge of the Miami Archipelago. This reef tract marks the initiation of the deposition that would result in the formation of the Florida Keys. At the northern end of the Palm Beach Archipelago, a large chain of sand and mangrove islands, the **Tomeu Islands** (named for the Tomeu family of Palm Beach Aggregates, Inc.) was beginning to form, greatly narrowing the **Loxahatchee Strait**. Only the **Caloosahatchee Strait** remained deep and wide at this time, and connected to a broad, shallow area of coastal lagoons to the north, the **Bermont Lagoon System** (named for the Bermont Formation).

The environments associated with these paleogeographic features produced the varied sediment types and lithologies seen in the lower member of the Bermont Formation. Before the formal description of the Holey Land Member, we here give a description and definition of the entire formation.

THE BERMONT FORMATION, OKEECHOBEE GROUP

HISTORY OF DISCOVERY

In the early 1960s, when quarries were being dug in the vicinity of Belle Glade, Palm Beach County, Druid Wilson first recognized a lithologically distinct set of beds containing a new and unstudied molluscan fauna. Wilson never published his discoveries, but Axel Olsson later gave an overview of his research (Olsson in Olsson and Petit, 1964: 521–522), naming the set of beds "Unit A." This unnamed formation was also referred to as the "Glades Unit" and its molluscan fauna was reported on by Thomas McGinty (1970), as well as by Hoerle (1970) with a detailed species list but now-outdated taxonomy). Finally, in 1974, Jules DuBar named the Unit A-Glades Unit beds the Bermont Formation, for the Bermont quadrangle (near the hamlet of Bermont, Charlotte County). DuBar's new formation was described from his "Unit F" along the Charlotte County section of Shell Creek, a tributary of the Peace River. Based on the distribution of some key index fossils, DuBar was able to show that his Bermont Formation was widely distributed around southern Florida (DuBar, 1974: Table 8).

During the late 1980s and early 1990s, new quarry excavations and canal dredgings in Dade, Broward, and Palm Beach Counties exposed some of the thickest and best-developed beds of the Bermont Formation that had ever been seen previously. Closer examination of these exposures showed that DuBar's formation was actually far more complex than originally described, with a far greater range of lithologies and greater stratigraphic development. Likewise, the molluscan fauna of the Bermont Formation was also found to be far richer, containing more endemic taxa than originally thought (Petuch, 1988, 1989, 1991, 1994, 2004). Working in quarries in western Miami (Capeletti Brothers Pit #11) and southwestern Palm Beach County (Griffin Brothers Pit), the senior author (Petuch, 1991) also found that the Bermont Formation comprised at least two distinct members, each containing its own distinctive molluscan fauna.

The Bermont Formation was also recognized as occurring in the Leisey Shell Corporation quarries in southern Hillsborough County and was reported on by Morgan and Hulbert, MacFadden, and Portell et al. in a special bulletin of the Florida Museum of Natural History (1995). Most recently, the paleoceanography and paleoecology of the Bermont Formation were described by the senior author (Petuch, 2004: 227–243). In the collection of review papers entitled *The Geology of Florida* (Randazzo and Jones, 1997), the Bermont Formation is missing from the generalized stratigraphic column shown on the inside covers. This omission is confusing, because Jones (115–116, Figure 7.22) later discusses the Bermont Formation and illustrates Bermont index fossils (from the "Pleistocene Bermont Formation of southern Florida").

LITHOLOGIC DESCRIPTION AND AREAL EXTENT

Condensed here is an extract from DuBar (1974: 221–222):

> The formation has been traced along Shell Creek for a distance of about seven miles. The unit typically is a gray, unconsolidated sandy shell marl. The sand fraction is composed of clear, subrounded, medium-grained quartz with a low percentage of admixed rounded, lustrous black phosphorite grains. Fossils are abundant and well preserved, and assemblages are quantitatively and qualitatively homogeneous throughout the area … An apparent facies of the Bermont Formation occurs in Palm Beach County along the southeast margin of Lake Okeechobee between Belle Glade and Port Mayaca … the Belle Glade facies is a gray shell marl.

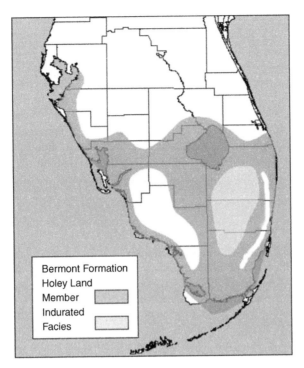

FIGURE 6.3 Areal distribution of the Holey Land Member of the Bermont Formation, showing the area containing heavily indurated limestone.

Morgan and Hulbert (1995: 21) described the Bermont beds exposed in the Leisey Shell pits as "1.0 m, dark, sandy mud; freshwater shells; bones" (lower bed) and "2.3 m, cream to bluish sandy, massive shell bed" (upper bed). The formation description is here expanded to include thick beds of heavily indurated fossiliferous limestones (biomicrudites and biocrudites, both marine and freshwater), highly arenaceous limestones (and limy sandstones), often containing petrified wood, unconsolidated yellow-tan quartz sands with scattered molluscan bioclasts, and lime muds (calcilutites) composed mainly of charaphyte phytoliths and packed with freshwater snails. The freshwater and marine limestone units are typically colored a light gray, whereas the sandy limestones are colored either golden-yellow or cream. Some limy sandstones are gray-brown to nearly black with dark gray or black silicified mangrove wood or dark gray silicified Turtle Grass (*Thalassia*) root casts.

With an expanded lithologic description and a larger and better-defined set of index fossils, the Bermont Formation is now known to be much more widely distributed than originally thought. As shown in Figure 6.3 and Figure 6.13 (with the two most widespread members), the Bermont Formation extends from southern Hillsborough County south to Monroe County and from southern Highlands and Okeechobee Counties south to Dade County. DuBar (1974: 221) states that the Bermont Formation extends northward up both coasts of Florida, to Levy County in the west and to Putnam County in the east. On the west coast, the beds north of southern Tampa Bay appear to be younger units, equivalent to the Fort Thompson Formation and do not belong in the Bermont Formation. Likewise, on the east coast, the beds north of Martin County appear to belong to the older Nashua Formation and also do not belong in the Bermont. Beach deposits above these Nashua beds, containing mostly quartz sand, may be equivalent to the Bermont. At the Rucks Pit in Fort Drum, Okeechobee County, the Bermont is represented by a thin (0.8 m) reddish-brown peaty sand layer, devoid of fossils, between the top of the Rucks Pit Member of the Nashua Formation and the base of the shell beds of the Coffee Mill Hammock Member of the Fort Thompson Formation.

Throughout its range, the Bermont is variable in thickness, being 3 m thick at the stratotype on Shell Creek, 10 m thick at Belle Glade, Palm Beach County, 4.5 m thick at the Leisey Shell pits, Hillsborough County, 12 m thick at Loxahatchee, Palm Beach County (Palm Beach Aggregates, Inc., quarries), and 15 m thick at the Capeletti Brothers Pit in Miami, Dade County. The surface contours are also variable, being 6 m below surface at the Leisey Shell pits, 5 m below Palm Beach Aggregates quarries, and 4 m below surface at the Griffin Brothers Pit in the Holey Land area. In the Everglades Agricultural Area (EAA), south of State Road 80 and between Belle Glade and the 20-Mile Bend at Loxahatchee, the surface of the Bermont undulates, ranging from 1.5 m to 5 m below land surface. These undulations are oriented in a northeast–southwest trend, are regularly spaced, and probably represent elongated sand bars and tidal channels that formed in the Loxahatchee Subsea. Within the Everglades Basin, the Bermont is the principal near-surface geologic formation.

STRATOTYPE

To quote from DuBar (1974: 221): " ... the name Bermont Formation is informally proposed here for the uppermost fossiliferous marine sands exposed along Shell Creek in the Bermont quadrangle, Charlotte County." The Bermont quadrangle is centered on the hamlet of Bermont, at the intersection of State Road 31 and State Road 74 (Bermont Road). (The only building at this intersection, the old Bermont General Store, burned down in 1998.) Shell Creek is a small tributary of the Peace River, flowing into it just north of its confluence with Charlotte Harbor. The cypress heads and marshlands around Bermont house the headwaters of Shell Creek; DuBar (1974: 221) stated that the "formation has been traced along Shell Creek for a distance of about seven miles." This is considered the stratotype area.

AGE AND CORRELATIVE UNITS

Based on paleomagnetic studies done at the Leisey Pit in Hillsborough County (MacFadden, 1995), the basal beds (Holey Land Member) date from around 1.4 to 1.6 million years B.P. (Aftonian Stage). Using uranium isotope ratio dating, the upper Bermont beds at Belle Glade (Belle Glade Member) were shown to date from approximately 750,000 to 800,000 years B.P. (Yarmouthian Stage) (Lyons, 1991: 156–157). From these data, the Bermont Formation is now known to span almost 1 million years and is correlative with the Flanners Beach Formation of the Carolinas.

BERMONT INDEX FOSSILS

Because the formation represents the deposition of two subseas and a paleolake, three sets of index fossils are recognized. Only a few species occurred in both the Loxahatchee and Belle Glade Subseas and only a few are found in both the Holey Land and Belle Glade Members. Some of these long-lived taxa include:

Gastropoda
Vokesimurex anniae
Vokesinotus griffini
Eupleura longior
Fasciolaria okeechobeensis
Latirus jucundus
Vasum floridanum (Figure 6.7H)
Bivalvia
Noetia platyura
Caloosarca aequilitas (Figure 6.14G)

The faunas of the Holey Land and Belle Glade Members, for the most part, contain very different-looking molluscan assemblages, each with a very high percentage of endemism. The freshwater molluscan fauna of Lake Okeelanta (Okeelanta Member) is one of the most bizarre ever discovered and easily compares with the African Rift Valley lakes for diversity and species radiations. These will be discussed under the descriptions of the individual members in the following sections.

THE HOLEY LAND MEMBER, BERMONT FORMATION

History of Discovery

In 1983, the senior author and his wife, while collecting at the Griffin Brothers quarries on the eastern edge of the Holey Land Wildlife Management Area of Palm Beach County, discovered a previously unknown lithofacies of the Bermont Formation. This unit, which was later informally named the "Holey Land Limestone" (Petuch, 1990) (Figure 6.4), was found to contain a highly endemic molluscan fauna, with many unusual taxa that had never been reported before. Besides containing classic Bermont index species such as *Miltha carmenae* and *Fasciolaria okeechobeensis*, the Holey Land Limestone also contained such oddities as the giant red-spotted olive shell *Lindoliva* and the siphonate melongenid subgenus *Miccosukea*. In 1986, the senior author encountered a Holey Land molluscan fauna in the Capeletti Brothers Pit #11, west of Miami, complete with *Lindoliva* and *Miccosukea* species. Unlike the Holey Land Limestone, these fossils were embedded in an unconsolidated sandy calcilutite (lime mud) and were exceptionally well preserved, retaining their original gloss and color patterns. (As a note to readers interested in Florida history, the area adjacent to the Griffin quarries was given the name "Holey Land" because it is peppered with shallow bomb craters, the result of having been an aerial bombing practice range during World War II. The bombers flew out of the Boca Raton Air Field, which is now our university, Florida Atlantic; the old landing strips are now the students' very long parking lots.)

On the northwestern side of the Holey Land, a new canal and levee system was excavated and built in 1999, giving access to new exposures of the Holey Land Limestone. Some of the canal areas were diked off (such as at the Talisman Canal) and later dewatered, allowing for the first *in situ* collecting of Bermont strata anywhere in the central Everglades. Here, as in the Griffin pits on the eastern side of the Holey Land, the thick limestone unit was seen to be overlain by a layer of unconsolidated sediments filled with classic Bermont fossils such as those seen at Belle Glade. A layer of freshwater calcarenite, filled with planorbid snails, was also present between the limestone and the Belle Glade-type unit. These canal sections substantiated the senior author's (Petuch, 1988, 1990) previous idea that the Bermont was composed of at least two separate members and was not a single unit.

In 1996, the senior author discovered another set of Bermont exposures in the Palm Beach Aggregates, Inc., quarries at Loxahatchee, Palm Beach County, west of West Palm Beach (previously known as the GKK Pit) (Figure 6.5 and Figure 6.6). At this locality, all three main units of the Bermont Formation were well developed and the lower set of beds was found to contain the same molluscan fauna as that of the Holey Land Limestone. The genus *Lindoliva* was present here, but rare, and other normally uncommon lower Bermont taxa were abundant. This indicates that a different set of depositional environments and ecosystems were present in the Loxahatchee area during Aftonian times. The lower beds of the Bermont at Palm Beach Aggregates also differed lithologically from those in Miami and the Holey Land area, consisting of fossiliferous sandy limestone units alternating with thin unconsolidated shell beds in a matrix of sandy lime mud (Figure 6.5). This distinctive unit was informally referred to as the "Loxahatchee Unit" or Member by the senior author (Petuch, 2004: 18). The lower Loxahatchee beds were lithologically intermediate between the unconsolidated sandy calcilutites of the Capeletti Brothers Pit and the massive indurated fossiliferous limestone of the Griffin Brothers Pit, forming a continuum.

FIGURE 6.4 Detail of a block of the heavily indurated facies of the Holey Land Member of the Bermont Formation, showing an embedded specimen of the cowrie shell *Macrocypraea spengleri* (Petuch, 1990). Along with the cowrie are (to its upper left) the remnants of a large rose coral (*Manicina*) and (below) numerous bivalves in the genera *Chione*, *Codakia*, and *Anadara*.

Lithologic Description and Areal Extent

We here give an expanded definition for the senior author's (Petuch, 1990) original informal description for the Holey Land and its lateral equivalents. The Holey Land Member typically consists of sandy (quartz), highly fossiliferous limestones (arenaceous biomicrudites and arenaceous crudites), often occurring in thick, heavily indurated units, beds of unconsolidated sandy (quartz) lime muds (arenaceous calcilutites) with densely packed molluscan bioclasts (primarily bivalves), and semifriable limy sandstones with high concentrations of organic matter (peats), containing silicified mangrove wood and Turtle Grass rhizomes. The fossiliferous limestones (Figure 6.4) are most commonly light gray to cream-gray in color, with white molluscan bioclasts (such as at the Holey Land stratotype), but can range in color from dark gray to grayish-brown (such as at the Palm Beach Aggregates quarries). The organic-rich, fine-particulate limy sandstones are generally dark

FIGURE 6.5 View of the eastern wall of the Palm Beach Aggregates, Inc. Pit #6 at Loxahatchee, Palm Beach County, showing a complete exposure of the Bermont Formation. Here, all three members are exposed, including the lower Holey Land Member (HL), the middle Okeelanta Member (O), and the upper Belle Glade Member (BG). The Holey Land Member can be seen to be lithologically complex, comprising interbedded units of sandy fossiliferous limestones and unconsolidated fossiliferous sandy calcarenites. The Holey Land extends below water surface for another 8 m. At this locality, the Okeelanta Member is thin and poorly developed, comprising an organic-rich sandy freshwater calcarenite filled with large freshwater snails. This unit was deposited in the shoreline areas (Marshland Zone) of Pleistocene Lake Okeelanta. The Belle Glade Member comprises a friable golden yellow sandy limestone filled with mollusk shells ("golden rock"). Photo by Anton Oleinik.

brown, but can vary from brownish gray to black. Petrified mangrove wood, if present (Petuch, 2004: plate 84), is usually colored a dark gray or grayish-brown. The unconsolidated sandy lime muds (arenaceous calcilutites and calcarenites) vary in color from yellow tan to light blue-gray (at the Capeletti Brothers Pit), to bluish (at the Leisey Pit), to dark brownish-gray (at Palm Beach Aggregates quarries) (Figure 6.5). The dense aggregations of molluscan fossils contained in these unconsolidated units vary from a distinctive blue-gray color (at Capeletti Brothers Pit #11) to bluish-white and pale tan (at Palm Beach Aggregates).

Under the northern and central Everglades of southwestern Palm Beach County, western Broward County, and northwestern Dade County (Figure 6.3), the Holey Land is completely indurated into a very hard (possibly the hardest limestone in southern Florida), massive fossiliferous limestone unit. To the north and east of the indurated area in central Palm Beach County, the Holey Land contains a mixed lithology, with alternating thin beds of fossiliferous limestones and unconsolidated shell beds in a fine quartz sand and calcilutite matrix (the "Loxahatchee Unit": Petuch, 1994) (Figure 6.6). Elsewhere, from Hillsborough County to Dade County, the Holey Land is composed almost com-

FIGURE 6.6 Detail of a bed of the strombid gastropod *Strombus evergladesensis* Petuch, 1991, in the Palm Beach Aggregates Pit #6 at Loxahatchee, Palm Beach County. This aggregation accumulated at the base of the Belle Glade Member of the Bermont Formation (the fossiliferous limestone unit above the strombids). (Photo by Anton Oleinik.)

pletely of unconsolidated sandy calcarenites and calcilutites, often containing dense shell beds. The shell beds are most commonly composed of bivalves, primarily a subspecies of the small venerid *Chione elevata* (previously referred to as *C. cancellata*, a related species from the West Indies), but may be composed of strombid gastropods (*Strombus*) (Figure 6.6) or *Pyrazisinus* potamidid gastropods. Some facies, such as those found in the Capeletti and Palm Beach Aggregates quarries, contain rich assemblages of corals, often with over 20 species being present (Petuch, 2004: 235–236). At the Griffin Brothers Pit, the bedding plane surface of the massive limestone was frequently covered with a pavement of *Manicina* rose corals and *Clypeaster* sea biscuits.

The Holey Land Member has the same areal distribution as does the entire Bermont Formation (Figure 6.3), extending from southern Hillsborough County to Monroe County and from southern Okeechobee County to Dade County. The member is absent under the Immokalee Rise and Atlantic Coastal Ridge, as these were islands during the time of deposition. Here, the Holey Land is replaced with undifferentiated sands. Throughout its range, the Holey Land Member varies in thickness from 3.5 m at the Leisey Shell pit, to over 6 m at the Griffin Brothers Pit, to 10 m at Palm Beach Aggregates, to over 15 m at the Capeletti Brothers Pit #11. The heavily indurated area of the Holey Land (Figure 6.3) roughly corresponds to the basin of Lake Okeelanta (described next). These massive limestones may have formed diagenetically through exposure in the deepwater areas of the paleolake. Here, the Holey Land sediments would have constituted the exposed floor of Lake Okeelanta, making them vulnerable to infiltration and recrystalization. Holey Land sediments covered by the peripheral marshland sediments of Lake Okeelanta, on the other hand, were protected from extensive diagenesis and remained unconsolidated.

Stratotype

Retaining the name and original type locality, we here designate the stratotype of the Holey Land Member to be the western side of the Griffin Brothers Pit #2, directly adjacent to the Holey Land Wildlife Management Area, on the power line access road, 7 km west of Deem City, at the intersection of Highway 27, Palm Beach–Broward County line, Palm Beach County, Florida. The Griffin pits are now flooded and have been incorporated into the Everglades Restoration Plan, as part of the filter marsh system. The Holey Land stratotype lies 2.5 m below mean water surface and can be accessed by scuba diving. (Deem City, once a hamburger shop and trailer, is now gone).

Age and Correlative Units

Based on uranium isotope dating (discussed earlier in this chapter) and on the presence of key widespread index fossils such as the bivalve *Noetia platyura*, the Holey Land Member is now known to date from the Aftonian Interglacial Stage of the early middle Pleistocene. The member is correlative with the lower beds of the Flanners Beach Formation of the Carolinas.

Holey Land Index Fossils

The Holey Land Member, including all its facies, contains a rich and highly endemic molluscan fauna. These are made up of three main components: widespread species found throughout the extent of the Holey Land, species found only within the central area of the pseudoatoll (Loxahatchee Trough), and relictual species that represent Kissimmean genera that survived into the Aftonian Pleistocene. Some of the more common and indicative species include the following, all illustrated and described by Petuch (1994):

Widespread Holey Land species
Gastropoda
Strombus erici (Figure 6.8D) (often misidentified as *S. alatus*)
Macrostrombus mayacensis (Figure 6.8G)
Fasciolaria okeechobeensis
Vasum floridanum (Figure 6.7H)

Species restricted to the Everglades area
Gastropoda
Astraea (Lithopoma) lindae
Cerithidea duerri
Pyrazisinus palmbeachensis (Figure 6.9D)
Pyrazisinus roseae (Figure 6.8F)
Pyrazisinus turriculus (Figure 6.9G)
Trivia (Niveria) bermontiana
Luria voleki (Figure 6.8A)
Macrocypraea joanneae (Figure 6.8H)
Macrocypraea spengleri (Figure 6.7E)
Pseudozonaria portelli (Figure 6.9H)
Macrostrombus diegelae
Macrostrombus scotti (Figure 6.7I)
Eustrombus gigas pahayokee (Figure 6.9F)
Titanostrombus williamsi (Figure 6.7J)
Cassis jameshoubricki (Figure 6.8I)
Cassis schnireli (Figure 6.7C)
Phalium loxahatcheensis

Vokesimurex diegelae
Fasciolaria (Cinctura) capelettii
Fasciolaria (Cinctura) holeylandica
Fusinus capelettii
Melongena (Miccosukea) cynthiae (Figure 6.9A)
Melongena (Miccosukea) holeylandica (Figure 6.9I)
Fulguropsis feldmanni
Fulguropsis capelettii
Sinistrofulgur holeylandicum (Figure 6.8K)
Microcythara caloosahatcheensis
Scaphella capelettii
Scaphella seminole
Lindoliva griffini (Figure 6.8E)
Lindoliva spengleri (Figure 6.7D)
Oliva (Strephona) adami
Oliva (Strephona) ryani
Oliva (Strephona) smithorum
Olva (Strephona) wendyae
Cariboconus griffini (Figure 6.9K)
Gradiconus capelettii (Figure 6.9B)
Gradiconus loxahatcheensis (Figure 6.9L)
Jaspidiconus hyshugari
Jaspidiconus maureenae
Bivalvia
Caloosarca catasarca

Relictual species (members of groups surviving from Caloosahatchee time)
Gastropoda
Pusula lindajoyceae (Figure 6.8C)
Pusula dadeensis
Jenneria hepleri
Jenneria loxahatchiensis (Figure 6.7B)
Malea petiti (Figure 6.7F)
Pyruella tomeui (Figure 6.7A)
Oliva (Porphyria) gravesae (Figure 6.9E)
Calusaconus tomeui (Figure 6.8J)
Ximeniconus palmbeachensis
Bivalvia
Conradostrea unnamed species, incorrectly referred to as "*C. sculpturata*" by Portell et al.
  (1995)
Carolinapecten jamieae (Figure 6.7G), incorrectly referred to as "*C. solaroides*" by Portell
  et al. (1995)
Armamiltha unnamed species (*A. disciformis* complex)
Miltha carmenae (Figure 6.9J)
Semele unnamed species (*S. perlamellosa* complex), incorrectly referred to as "*S. perlamel-
  losa*" by Portell et al. (1995)

All these relictual genera, with the exception of *Calusaconus* and *Ximeniconus*, were extinct
or extirpated in Florida by the end of Holey Land time, and do not occur in the younger Belle
Glade Member. Other classic Holey Land invertebrate fossils include the delicate branching coral

FIGURE 6.7 Index fossils for the Holey Land Member of the Bermont Formation. A = *Pyruella tomeui* Petuch, 2004, length 89 mm; B = *Jenneria loxahatchiensis* M. Smith, 1936, length 29 mm; C = *Cassis schnireli* Petuch, 1994, length 221 mm; D = *Lindoliva spengleri* Petuch, 1988, length 110 mm; E = *Macrocypraea spengleri* (Petuch, 1990), length 121 mm; F = *Malea petiti* Petuch, 1989, length 122 mm; G = *Carolinapecten jamieae* Petuch, 2004, length 94 mm; H = *Vasum floridanum* McGinty, 1940, length 117 mm; I = *Macrostrombus scotti* (Petuch, 1994), length 183 mm; J = *Titanostrombus williamsi* (Olsson and Petit, 1964), length 265 mm.

FIGURE 6.8 Index fossils for the Holey Land Member of the Bermont Formation. A = *Luria voleki* Petuch, 2004, length 38 mm; B = *Rhyncholampas* cf. *evergladesensis* Mansfield, 1932; C = *Pusula lindajoyceae* Petuch, 1994, length 22 mm; D = *Strombus erici* Petuch, 1994, length 95 mm; E = *Lindoliva griffini* Petuch, 1988, length 73 mm (lower of the Holey Land only, differs from *L. spengleri* in that it is a smaller, more inflated shell with a lower spire); F = *Pyrazisinus roseae* Petuch, 1991, length 58 mm; G = *Macrostrombus mayacensis* (Tucker and Wilson, 1933), length 148 mm; H = *Macrocypraea joanneae* Petuch, 2004, length 54 mm; I = *Cassis jameshoubricki* Petuch, 2004, length 183 mm; J = *Calusaconus tomeui* Petuch, 2004, length 34 mm; K = *Sinistrofulgur holeylandicum* Petuch, 1994, length 105 mm (differs from the Recent *S. sinistrum* in that it has a stronger spiral sculpture and more rounded body whorl).

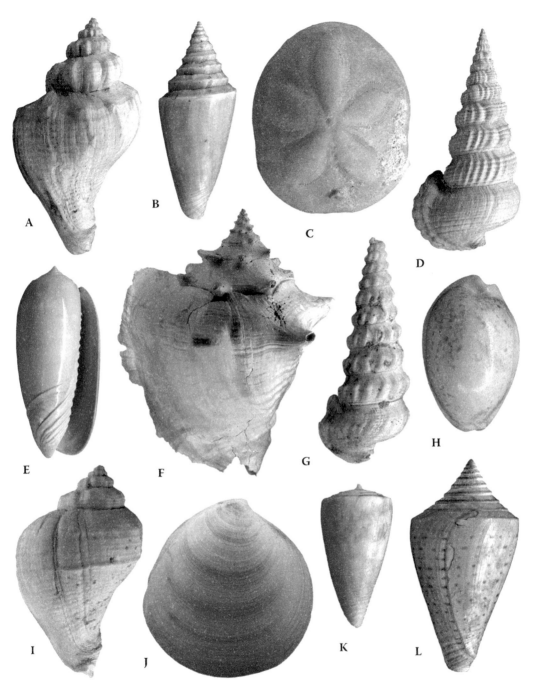

FIGURE 6.9 Index fossils for the Holey Land Member of the Bermont Formation. A = *Melongena (Miccosukea) cynthiae* Petuch, 1990, length 61 mm; B = *Gradiconus capelettii* (Petuch, 1990), length 42 mm; C = *Clypeaster rosaceus* subspecies, length 91 mm; D = *Pyrazisinus palmbeachensis* Petuch, 1994, length 62 mm; E = *Oliva (Porphyria) gravesae* Petuch, 1994, length 57 mm; F = *Eustrombus gigas pahayokee* (Petuch, 1994), length 179 mm; G = *Pyrazisinus turriculus* Petuch, 1994, length 58 mm; H = *Pseudozonaria portelli* (Petuch, 1990), length 23 mm (note speckled color pattern); I = *Melongena (Miccosukea) holeylandica* Petuch, 1990, length 59 mm; J = *Miltha carmenae* H.Vokes, 1969, length 82 mm; K = *Cariboconus griffini* (Petuch, 1990), length 18 mm; L = *Gradiconus loxahatcheensis* (Petuch, 1994), length 51 mm.

Arcohelia limonensis, the echinoid echinoderms *Rhyncholampas* cf. *evergladesensis* (Figure 6.8B), and *Clypeaster rosaceus* (Figure 6.9C). The marine communities of the Holey Land Member were described by the senior author (Petuch, 2004: 228–236) and include the *Pyrazisinus roseae* Community (mangrove forests and intertidal mud flats), the *Titanostrombus williamsi* Community (Turtle Grass beds), the *Carolinapecten jamieae* Community (shallow sand-bottom lagoons), and the *Arcohelia limonensis* Community (coral reefs and coral bioherms). The impoverished nature of the Leisey Shell Pit Holey Land fauna, lacking these tropical communities, demonstrates that the Aftonian Gulf of Mexico had cooler water conditions than did the enclosed Everglades basin.

THE OKEELANTA MEMBER, BERMONT FORMATION

History of Discovery

From 1980 until 1994, a section of the North New River Canal, along State Highway 27, 3–5 km south of South Bay, Palm Beach County (adjacent to the Okeelanta Corp. sugar mill property), was regularly dredged and widened. During these excavations, large blocks of a highly fossiliferous pale orangish-tan calcarenite were brought up by the draglines and tossed onto the top of the canal berm (capped with a dirt access road). The senior author visited this site on a continuous basis over the time of the excavations, and discovered that these "marl" blocks were filled with a highly unusual freshwater molluscan fauna. Of particular interest was a large number of species of planorbid gastropods (genera *Planorbella* and *Seminolina*), including bizarre elongated and uncoiled forms that had never before been reported. These distinctive index fossils were later found at other localities along the North New River Canal and Highway 27, from 5 to 20 km south of the Okeelanta site.

In the walls of the dewatered canals exposed during the construction of the Holey Land levee (discussed in the previous section), this freshwater calcarenite was found to be a much more complicated and thicker unit, also containing limestone layers, beds of freshwater clams, and chara-phyte phytolith lime muds (phytolith calcilutites). At the Holey Land levee site, these freshwater beds were subjacent to a sandy calcarenite filled with typical Belle Glade marine fossils. This same depositional pattern was seen during dragline dredgings in the Griffin Brothers Pit. The limestone units at both localities contained either densely packed assemblages of the tiny hydrobioid gastropod *Fontigens*, small *Seminolina* planorbid gastropods, or beds of unnamed species of *Corbicula* and freshwater *Rangia* clams. The Okeelanta unit is also present in the Palm Beach Aggregates quarries at Loxahatchee, but here it is poorly developed and forms a thin bed between the Holey Land and Belle Glade Members (Figure 6.5). At this locality, *Seminolina* snails are outnumbered by a large discoidal *Planorbella* species, *Viviparus* snails, *Stenophysa* species, and *Pomacea* apple snails.

Lithologic Description and Areal Extent

This middle Bermont unit, here informally named the Okeelanta Member, was deposited within a large Pleistocene freshwater body, Lake Okeelanta (Petuch, 2004: 236). This Everglades paleolake formed during the sea level low of the Kansan Glacial Stage (Figure 6.1). In its maximum development during this emergent time, Lake Okeelanta was a deep (at least 20 m), holomictic lake that encompassed an area approximately four times the size of Recent Lake Okeechobee (Figure 6.10). Occupying the basin of the Okeechoban Sea, Lake Okeelanta formed in a rich carbonate sedimentary environment, and its water was saturated with dissolved calcium carbonate. This saturation is reflected in the thick layers of charaphyte phytolith lime muds that were produced by calcareous green algae, such as *Chara*, which thrive in this type of water chemistry. Lake Okeelanta was analogous to an inland sea, producing the richest and most diverse lacustrine molluscan fauna known from anywhere in North America.

Based on the distribution of the Okeelantan molluscan assemblages and on the ecology of living related forms, three broad, interfingering ecological zones can be delineated within the paleolake boundaries: (1) a peripheral **Marshland Zone**, dominated by emergent vegetation and containing

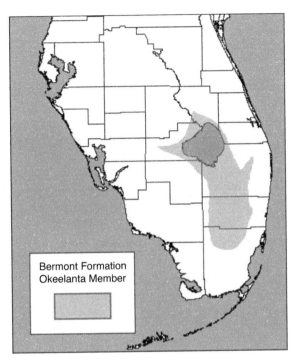

FIGURE 6.10 Areal distribution of the Okeelanta Member of the Bermont Formation, following the outline of Pleistocene Lake Okeelanta.

a molluscan fauna made up of *Viviparus*, *Stenophysa*, *Pomacea*, and large discoidal *Planorbella* snails; (2) a **Charaphyte Zone**, comprising shallow, open bottom areas that supported dense beds of the green calcareous alga *Chara*; and (3) a **Corbicula Zone**, with dense beds of *Corbicula* and *Rangia* clams living on open bottoms in deeper water. The bizarre, elongated *Seminolina* species shown in Figure 6.11 lived in the *Corbicula* Zone, and this assemblage represents the single most unusual freshwater community ever found in Florida. Each of the ecological zones was also associated with distinct types of deposition, ranging from muddy sands with high levels of organic matter to pure, clean carbonate muds.

The Okeelanta Member typically consists of interbedded layers of freshwater calcarenites, generally highly fossiliferous. These range from unconsolidated lime muds (calcilutites) packed with planorbid gastropod or corbiculid and mactrid bivalve bioclasts, charaphyte phytolithic calcarenites with planorbid gastropod bioclasts, sandy (quartz) lime muds with high concentrations of organic matter (organic-rich arenaceous calcilutites) and with large molluscan bioclasts, mostly *Viviparus*, *Pomacea*, and discoidal planorbids, and thin fine-grained limestone units (biomicrudites) with abundant molluscan bioclasts. The calcilutites vary in color from pale orange-tan, to light brown, light gray, or pure white. In these units, the shell bioclasts are usually stained tan, brown, light gray, or dark bluish-gray. The sandy lime muds are typically of a dark brown or brown-gray color and their molluscan bioclasts are stained a dark brown, dark gray, or black color. The thin limestone units, usually occurring as duricrusts on the bedding plane surface, are colored light brown or light gray and contain white molluscan bioclasts.

The Okeelanta Member has an areal extent that conforms to the outline of Lake Okeelanta and has the smallest distribution of the three Bermont members (Figure 6.10). Outside the boundaries of the paleolake, an unconformity or terrestrial sequence takes the place of the member. Okeelanta sediments are known only from Glades, southern Okeechobee, westernmost Martin, Palm Beach, northeasternmost Hendry, Broward, and northern Dade Counties. In the stratotype area at Okeelanta and in the Holey Land, the member averages 1.5–2 m in thickness (deposited in the deepest part of the lake). In the Palm Beach Aggregates quarries (Figure 6.5), the member was deposited in the peripheral shallow marshland areas and is only 0.5 m thick.

Stratotype

We designate a section on the eastern side of the North New River Canal, 5.2 km south of the city of South Bay, Palm Beach County, as the stratotype of the Okeelanta Member. The type locality is adjacent to the bridge at the intersection of State Highway 27 and County Road 827, northeast of the Okeelanta Corp. sugar mill property. The member crops out in the North New River Canal at a depth of 2.5 m below water surface. Specimens of Okeelanta fossils can be collected on the dirt access road on the top of the canal berm (dredged from the canal and used as road fill). The member takes its name from the Okeelanta sugar mill and the agricultural village of Okeelanta, which was destroyed in the 1928 hurricane.

Age and Correlative Units

Based on the stratigraphic position between the two uranium isotope-dated Bermont members, the Okeelanta Member is now known to date from the Kansan Glacial Stage of the middle Pleistocene.

Okeelanta Index Fossils

Present research indicates that the molluscan fauna of the Okeelanta Member is almost completely endemic, with only a few species known to occur outside the paleolake boundaries and with most species being restricted to the lake basin. Here, the different species groups were distributed by ecological preferences, and these assemblages can be used to determine the bathymetry of the paleolake. Only a few of the Okeelanta mollusks have been described to date (shown here), and at least 80 unnamed species remain to be described by future workers. Of these, the planorbid genus *Seminolina* exhibited the largest species radiation, with over 30 different species having been collected. The following are some of the more abundant and indicative Lake Okeelanta mollusks (arranged by ecological zone):

Corbicula Zone (freshwater bivalve beds)
Gastropoda
Seminolina knepperi (Figures 6.11C, D)
Seminolina cf. *knepperi* (Figures 6.11A, B)
Seminolina lindae (Figure 6.11H)
Seminolina cf. *lindae* (Figures 6.11F, G)
Fontigens unnamed species (elongated)
Elimia effosa
Bivalvia
Corbicula unnamed species
Rangia unnamed species (freshwater form similar to *R. lecontei* from the Pleistocene of California)

Charaphyte Zone (calcareous algal beds)
Gastropoda
Seminolina clewistonense (Figure 6.11M)
Seminolina roseae (Figures 6.11I, J)
Seminolina seminole subspecies (Figure 6.11K)
Seminolina wilsoni

FIGURE 6.11 Index fossils for the Okeelanta Member of the Bermont Formation. A,B = *Seminolina* cf. *knepperi* Petuch, 1991, length, 27 mm; C,D = *Seminolina knepperi* Petuch, 1991, length 15 mm (typical form); E = *Stenophysa barberi* (Clench, 1925), length 29 mm; F,G = *Seminolina* cf. *lindae* Petuch, 1991, length 29 mm; H = *Seminolina lindae* Petuch, 1991, length 28 mm (typical form); I,J = *Seminolina roseae* Petuch, 1991, length 16 mm; K = *Seminolina seminole* (Pilsbry, 1934) subspecies, length 16 mm; L = *Planorbella preglabrata* (Marshall, 1926), width 25 mm; M = *Seminolina clewistonense* (Baker, 1940), width 26 mm.

Amnicola unnamed species (at least three species)
Fontigens palaea
Fontigens cf. *plana*
Probithinella unnamed species (at least two species)
Notogillia unnamed species (at least three species)
Viviparus unnamed species (several elongated forms with stepped, scalariform spires)

Marshland Zone (emergent vegetation)
Gastropoda
Viviparus cf. *georgianus*
Pomacea paludosa subspecies
Pomacea innexa subspecies
Planorbella preglabrata (Figure 6.11L)
Stenophysa barberi (Figure 6.11E)
Pseudosuccinea aperta
Oxyloma barberi
Bivalvia
Elliptio unnamed species

The Okeelanta Member also commonly contains alligator teeth, alligator osteoderms, and freshwater turtle shell fragments (primarily in the Marshland Zone).

PALEOGEOGRAPHY OF THE BELLE GLADE SUBSEA, OKEECHOBEAN SEA

Around 700,000 years B.P., worldwide climates began to warm, the Kansan continental ice sheets melted, and sea level began to rise, ushering in the Yarmouthian Interglacial Stage (Figure 6.1). With the initiation of this new period of global warming, southern Florida was engulfed by a new marine system, the Belle Glade Subsea (named for the city of Belle Glade, Palm Beach County) (Figure 6.12). The rich endemic faunas of Kansan Lake Okeelanta quickly perished in the influx of seawater, with only a few species finding refuges in marshlands on higher elevation areas (the ancestors of the Recent Everglades freshwater mollusks). During the preceding Kansan time, a large percentage of the Loxahatchee Subsea molluscan fauna became extinct, leaving a much-diminished fauna to populate the ecosystems of the Belle Glade Subsea. Some of the more important groups that disappeared from the Florida fossil record during Kansan time included the gastropod genera *Pyruella*, *Malea*, and *Jenneria*, and the bivalve genera *Carolinapecten*, *Conradostrea*, and *Armamiltha*. The disappearance of these organisms, and the addition of newly evolved endemic species, gives the Belle Glade fauna a distinctive appearance that can be used as a stratigraphic marker.

The geomorphology of the Belle Glade Subsea was essentially an extension of the predecessor Loxahatchee Subsea, with the older features being enlarged and better developed during Yarmouthian time. **Immokalee Island** had now fused with mangrove forests to the western side of **Miccosukee Island**, obliterating the shallow lagoon and Turtle Grass environments of the Hendry Platform. **Cape Sable Island** had now fused to the southern section of the **Miami Archipelago**, forming a continuous chain of large, wide, pine-forested islands. Likewise, the **Palm Beach Archipelago** and the **Tomeu Islands** had widened by this time, effectively closing off the Atlantic from the Belle Glade Subsea. Only a few narrow cuts existed between the islands, and these remain today as the Transverse Glades low areas on the Atlantic Coastal Ridge of Palm Beach, Broward, and Dade Counties. Because of the deposition of Lake Okeelanta and its surrounding marshlands during Kansan times, the **Loxahatchee Trough** of the Belle Glade Subsea was much shallower

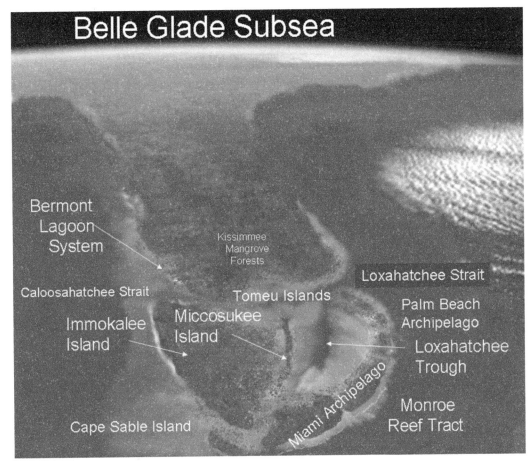

FIGURE 6.12 Simulated space shuttle image (altitude: 100 mi) of the Floridian Peninsula during the Yarmouthian Pleistocene, showing the possible appearance of the Belle Glade Subsea of the Okeechobean Sea and some of its principal geomorphologic features.

than it was in the Loxahatchee Subsea. During Yarmouthian times, only a small, narrow central area retained deepwater conditions, whereas the rest of the trough was composed of broad, shallow carbonate banks (the center of deposition for the Belle Glade Member). To the south, the **Monroe Reef Tract** was also becoming better developed and extended to at least present-day Bahia Honda Key in the Florida Keys.

Along the northern end of the Belle Glade Subsea, the **Bermont Lagoon System** was also rapidly filling with sediments during Yarmouthian times, supporting large cypress forests and coastal marshlands. This explains the diminished presence of Belle Glade Member beds at the Leisey Shell Pit in Hillsborough County. Mangrove forests filled the southern region of the old Kissimmee Embayment (the **Kissimmee Mangrove Forests**), whereas the northern end had now filled with hardwood and pine forests. This explains the absence of Bermont marine beds at the Rucks Pit at Fort Drum, Okeechobee County (see Chapter 5). The **Loxahatchee Strait** was filling with sediments at this time and contained large carbonate shoals and Turtle Grass beds. Only the **Caloosahatchee Strait** remained deep, being scoured by strong tidal currents that prevented the accumulation of any thick beds of sediments. This explains the absence, or extreme truncation, of Bermont beds along the present-day Caloosahatchee River. The environments contained in these geomorphologic features produced the lithofacies and biofacies of the last pulse of Bermont deposition, the Belle Glade Member.

THE BELLE GLADE MEMBER, BERMONT FORMATION

History of Discovery

Comprising the upper beds of the Bermont Formation, the Belle Glade Member (named for the city of Belle Glade, Palm Beach County) contained the first set of beds recognized as being mid-Pleistocene in age. As mentioned earlier in this chapter, Druid Wilson (Olsson and Petit, 1964) noticed that the molluscan fossils indicated that this unit fell somewhere between the late Pliocene–early Pleistocene Caloosahatchee Formation and the late Pleistocene Fort Thompson Formation. The geochronological placement of the "Glades Unit" and "Unit A" by these early workers, using only molluscan biostratigraphy, was amazingly accurate.

From 1964 to 1969, the first large-scale exposures of sediments from the Belle Glade Member were brought up by draglines excavating a road fill rock pit in the city of Belle Glade (Hoerle, 1970; McGinty, 1970). Shirley Hoerle collected and identified 434 species of mollusks from this quarry, attesting to the richness of this upper Bermont fauna (her taxonomy, however, is now outdated, with many inaccuracies). In the 1980s, the senior author encountered the same Belle Glade fauna and sediments during the deepening and widening of the North New River Canal, along Highway 27, 3 to 20 km south of South Bay, Palm Beach County (discussed in the previous section on the Okeelanta Member). While working in the Holey Land area, the senior author also encountered Belle Glade beds at the Griffin Brothers Pit and at the Talisman Canal levee excavations. Similar sediments and molluscan faunas were also found at the Palm Beach Aggregates quarries in Loxahatchee (Figure 6.5) and at the Capeletti Brothers quarries in Miami, demonstrating that this member is widely distributed throughout the Everglades area.

In 1978, Muriel Hunter was the first person to apply a semiformal name to the beds encountered within the Belle Glade rock pit and south of Lake Okeechobee (during dredgings of the Miami and North New River Canals). Considering these beds to be correlative with the Okaloacoochee Member, she named them the "Belle Glade Member" and placed her new member in the late Pleistocene Fort Thompson Formation. Subsequent research has shown that the Belle Glade beds are much older than originally presumed by Hunter and that, cyclostratigraphically and lithologically, they belong in the Bermont Formation. We here wish to retain Hunter's "Belle Glade Member" but place it in the chronostratigraphic position as the Yarmouthian component of Bermont deposition.

Lithologic Description and Areal Extent

Typically, the Belle Glade Member is composed of thick beds of fine-particulate unconsolidated calcarenites with variable amounts of quartz sand and scattered coral and molluscan bioclasts. Although some beds (such as those seen at Palm Beach Aggregates quarries and along the North New River Canal) contain abundant fossil mollusks, these never form the densely packed masses, with little matrix, seen in the underlying Holey Land Member. Heavily indurated limestones are also absent from this member, although some facies (such as at Palm Beach Aggregates) are partially indurated, forming a friable, weak, sandy limestone (referred to locally as "golden rock"). Some of the calcarenite facies, such as those from the North New River Canal south of South Bay, are very clean, with only a small amount of quartz sand, and contain a coarser-grained calcarenite, almost approaching an oölite in appearance. Some calcarenite beds, such as those exposed along the Talisman Canal on the northern edge of the Holey Land, are extremely fine grained, approaching a calcilutite. The sandy calcarenites vary in color from white, to pale gray, to a light yellow-orange, to dark yellow. Shells preserved in the white to pale gray calcarenites are stained a dark blue-gray or golden-tan color, whereas those in the yellow-orange sandy calcarenites ("golden rock") are white. Equivalent beds outside the Everglades Basin, in cooler water areas such as at the Bermont stratotype at Shell Creek, typically contain mud units and have a much higher percentage of quartz sand and molluscan bioclasts. These sandy shell beds are typically a gray or blue-gray color. Within the stratotype area in western Palm Beach County and at the Capeletti quarries in Miami, large

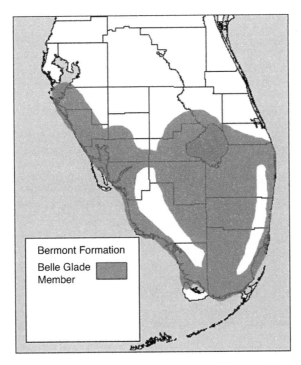

FIGURE 6.13 Areal distribution of the Belle Glade Member of the Bermont Formation.

beds of the branching coral *Porites furcata* are frequently encountered in the coarse-grained calcarenite beds. A thin (10 cm) laminar duricrust is present on the surface of some Belle Glade units.

The Belle Glade Member has the same areal distribution as does the older Holey Land Member, extending from southern Hillsborough County to Monroe County and southern Okeechobee County south to Dade County (Figure 6.13). Throughout this area, the member varies in thickness from 1–3 m at Shell Creek, Charlotte County to 10 m at the Belle Glade rock pit, Palm Beach County, to 3 m in the Palm Beach Aggregates quarries, Palm Beach County.

Stratotype

Retaining Hunter's (1978) original type locality, we here designate the stratotype of the Belle Glade Member to be the old Belle Glade rock pit, at the bend of State Road 80 adjacent to the Glades Memorial Hospital, Belle Glade, Palm Beach County (W ¹/₂, NE ¹/₄, Sec. 7, T. 44 S., R. 37 E., Palm Beach County). The quarry is now flooded as a lake, but is still accessible for collecting. The top of the Belle Glade Member crops out within the lake at a depth of 1.8 m below surface and can be accessed by scuba diving.

Age and Correlative Units

Based on the uranium isotope dates discussed earlier in this chapter, the Belle Glade Member is now known to date from the Yarmouthian Interglacial Stage of the late middle Pleistocene. The member is correlative with the upper beds of the Flanners Beach Formation of the Carolinas.

Belle Glade Index Fossils

Although containing a less diverse fauna than did the Holey Land Member, the Belle Glade still has rich and highly endemic molluscan assemblages that can be used to demarcate the member boundaries. The following are some of the more abundant and indicative Belle Glade species:

Gastropoda
Astraea (Lithopoma) southbayensis
Turbo (Marmarostoma) duerri
Calliostoma lindae
Pyrazisinus miamiensis (Figure 6.15B)
Strombus evergladesensis (Figure 6.14J)
Strombus lindae (Figure 6.14E)
Macrostrombus holeylandicus (Figure 6.14D)
Chicoreus duerri (Figure 6.15G)
Chicoreus gravesae
Vokesimurex bellegladeensis (Figure 6.14A) (still living in deep water off eastern Florida)
Vokesimurex bermontianus
Murexiella graceae (Figure 6.15D)
Fasciolaria (Cinctura) evergladesensis (Figure 6.14B)
Fusinus watermani (Figure 6.15K)
Melongena lindae (Figure 6.15A)
Melongena (Rexmela) bispinosa (Figure 6.14H) (still living along the Yucatan Peninsula)
Melongena (Rexmela) diegelae (Figure 6.15C)
Fulguropsis evergladesensis (Figure 6.15J)
Sinistrofulgur roseae (Figure 6.14C)
Eurypyrene miccosukee
Turbinella hoerlei (Figure 6.14K)
Lindoliva diegelae (Figure 6.15I)
Oliva (Strephona) cokyae
Oliva (Strephona) edwardsae (Figure 6.14I)
Oliva (Strephona) lindae (Figure 6.15F)
Oliva (Strephona) murielae
Oliva (Strephona) southbayensis
Calusaconus evergladesensis (Figure 6.14F)
Spuriconus lemoni (Figure 6.15H)
Spuriconus micanopy
Neodrillia blacki (still living in the Florida Keys)
Bivalvia
Caloosarca aequilitas (Figure 6.14G)
Nodipecten pernodosus (Figure 6.15E)

The marine communities of the Belle Glade Member were described by the senior author (Petuch, 2004: 236–243) and include the *Strombus lindae* Community (shallow carbonate banks), the *Caloosarca aequilitas* Community (Turtle Grass beds), and the *Nodipecten pernodosus* Community (shallow sand-bottom lagoons).

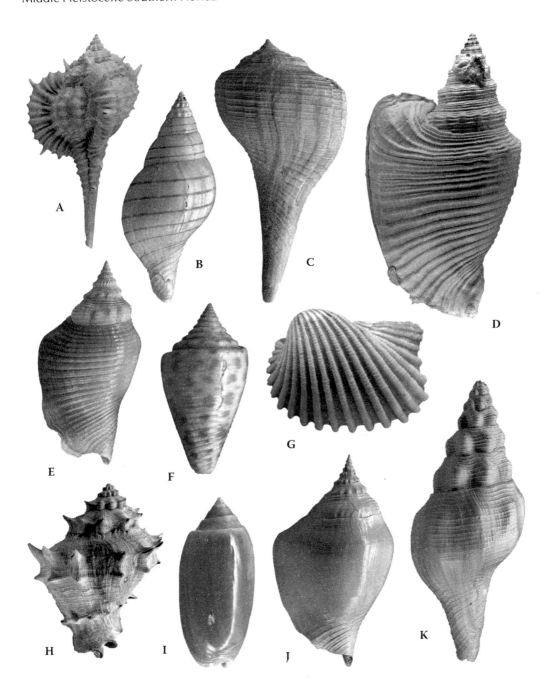

FIGURE 6.14 Index fossils for the Belle Glade Member of the Bermont Formation. A = *Vokesinotus bellegladeensis* (E.Vokes, 1963), length 47 mm; B = *Fasciolaria (Cinctura) evergladesensis* Petuch, 1991, length 47 mm; C = *Sinistrofulgur roseae* Petuch, 1991, length 109 mm (differs from the older Holey Land *S. holeylandicum* in that it has a much coarser and more pronounced spiral sculpture and a lower, flatter spire; this is the most heavily sculptured of all the left-handed whelks); D = *Macrostrombus holeylandicus* (Petuch, 1994), length 194 mm; E = *Strombus lindae* Petuch, 1991, length 44 mm; F = *Calusaconus evergladesensis* (Petuch, 1991), length 29 mm; G = *Caloosarca aequilitas* (Tucker and Wilson, 1932), length 56 mm; H = *Melongena (Rexmela) bispinosa* (Philippi, 1844), length 34 mm; I = *Oliva (Strephona) edwardsae* Olsson, 1967, length 30 mm; J = *Strombus evergladesensis* Petuch, 1991, length 71 mm; K = *Turbinella hoerlei* E.Vokes (1966).

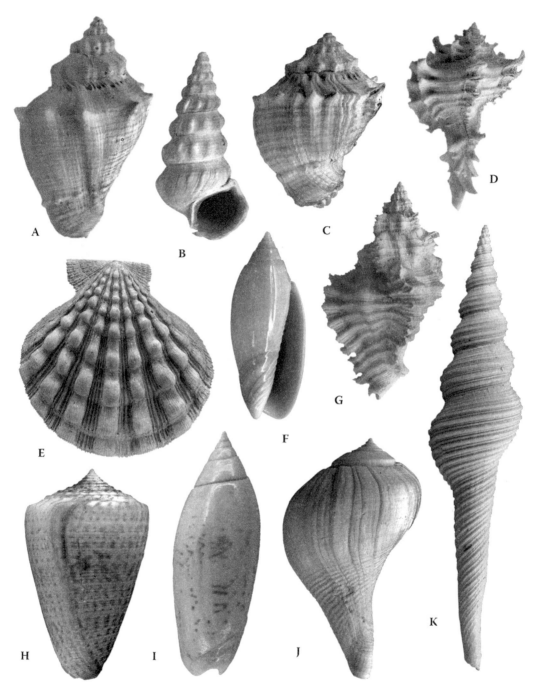

FIGURE 6.15 Index fossils for the Belle Glade Member of the Bermont Formation. A = *Melongena lindae* Petuch, 1994, length 45 mm; B = *Pyrazisinus miamiensis* Petuch, 1994, length 33 mm; C = *Melongena (Rexmela) diegelae* Petuch, 1994, length 27 mm; D = *Murexiella graceae* (McGinty, 1940), length 28 mm; E = *Nodipecten pernodosus* (Heilprin, 1886), length 98 mm; F = *Oliva (Strephona) lindae* Petuch, 1991, length 28 mm; G = *Chicoreus duerri* Petuch, 1994, length 32 mm; H = *Spuriconus lemoni* (Petuch, 1990), length 57 mm; I = *Lindoliva diegelae* Petuch, 1988, length 51 mm (smallest species of *Lindoliva*; differs from the Holey Land *L. spengleri* in that it is smaller and more slender); J = *Fulguropsis evergladesensis* Petuch, 1994, length 51 mm; K = *Fusinus watermani* (M. Smith, 1940), length 128 mm.

7 Late Pleistocene to Holocene Southern Florida

After a 400,000-year period of warm, subtropical conditions within the Belle Glade Subsea, sea levels dropped precipitously, and the entire Florida Platform was again emergent. Caused by the buildup of immense continental glaciers around the Northern Hemisphere, this sea level drop marks the beginning of both the Illinoian Glacial Stage and the late Pleistocene. Like many previous glacial times, the Illinoian was complex, with at least two glacial pulses separated by a short, warmer interglacial time (see Figure 6.1, Chapter 6). As severe as the first Illinoian glaciation was, the second was even more devastating, the sea level dropping to the lowest it had been since the Oligocene. Much of the rich Belle Glade marine fauna disappeared at this time.

As during the Kansan Glacial Stage, the Loxahatchee Trough area was flooded with freshwater during Illinoian time. Over the period of Belle Glade deposition, marine sedimentation filled many of the deeper areas within the Loxahatchee Trough, producing a broader, wider, and shallower basin. Because of this Yarmouthian sedimentary filling, no single deep lake existed during Illinoian time. Instead, the Loxahatchee Trough area was covered with a system of interconnected ephemeral marshlands and small lakes, not unlike the Recent Everglades. Any peat units put down by these early Illinoian paludal environments, however, would have been scoured away by subsequent marine incursions.

By the middle of Illinoian time, approximately 300,000 years B.P., a brief sea level rise occurred, flooding the Okeechobean Sea basin with very shallow marine and estuarine environments. This minor sea level rise marked the beginning of the Lake Worth Subsea (named for the city of Lake Worth, Palm Beach County; see Petuch, 2004), the last subdivision of the Okeechobean Sea (Figure 7.1). Within 100,000 years of being inundated with marine environments, the sea withdrew and southern Florida was again exposed to subaerial, terrestrial conditions. This occurred during the second, and most severe, of the Illinoian glacial pulses (Figure 6.1, Chapter 6). Throughout this time, Florida experienced dry, semiarid conditions and the Everglades area more closely resembled east central Texas than it does the present-day marshlands. A large, shallow freshwater body, Lake Pahayokee (see Petuch, 2004: 247), formed in the northern part of the Everglades Basin but lasted for only a short period of a few thousand years. This last great paleolake was populated with the descendants of the survivors from Lake Okeelanta.

After the lowest sea level drop of the Neogene period, the worldwide oceanic system quickly rebounded and flooded the continental margins with the highest known sea level, possibly 20 m above that of the present. This global warming-induced eustatic high at 150,000 years B.P. (Figure 6.1, Chapter 6) ushered in the Sangamonian Interglacial Stage and the second, and greatest, development of the Lake Worth Subsea. Together, the Sangamonian and mid-Illinoian sea level rises produced the environments for the deposition of the last extensive Everglades geologic unit, the Fort Thompson Formation (with its mid-Illinoian Okaloacoochee and Sangamonian Coffee Mill Hammock Members). The varied environments of the Sangamonian Lake Worth Subsea also produced the Key Largo Formation (coral reef tracts), the Miami Formation (oölitic banks), and the Anastasia Formation (beach deposits). The Sangamonian Lake Worth Subsea contained the last recognizable remnants of the Everglades Pseudoatoll.

At 75,000 years B.P., worldwide climates again began to cool and sea levels dropped during the initiation of the Wisconsinan I Glacial Stage. The Lake Worth Subsea quickly drained off the Everglades area, and southern Florida again was covered by semiarid terrestrial environments. The area of Illinoian Lake Pahayokee and Recent Lake Okeechobee was again flooded by rivers flowing off the Florida highlands, producing a series of small, shallow, ephemeral lakes. Freshwater

FIGURE 7.1 Simulated space shuttle image (altitude: 100 mi) of the Floridian Peninsula during the Sanga-monian Pleistocene, showing the possible appearance of the Lake Worth Subsea of the Okeechobean Sea and its principal geomorphologic features.

calcarenites deposited within these shallow lakes produced the Lake Flirt Formation (named for Lake Flirt, the latest Wisconsinan–Holocene descendant of this lake system). At the end of Wisconsinan II time, around 20,000 years B.P., sea levels again began to rise, and large transverse sand dunes formed along the eastern coast of Florida. These southwardly migrating quartz sands, often referred to as the "Pamlico Formation," began to cover the old pseudoatoll islands and were blown inland to form the substrate for the coastal pine forests.

Within the Everglades area, the subaerially hardened freshwater calcarenites of Lakes Pahayokee and Flirt acted as impervious confining units, pooling the water draining from the Kissimmee River and producing a broad, shallow sheet flow. By 7000 years B.P., in the early Holocene, the saw grass growing in the sheet flow area had begun to accumulate the thick peat layers that we associate with the present-day Everglades. As the world climate warmed at this time, increased rainfall in Florida would have fed and expanded the Everglades sheet flow, enlarging Lakes Okeechobee, Hicpochee, and Flirt, and producing the extensive cypress and pond apple forests that surround the saw grass prairies. All the environments that we associate with modern southern Florida, the Everglades marshlands and cypress forests, the coastal sand dunes and pine forests, and the peripheral mangrove jungles, rose from thin depositional films that formed during Holocene times.

PALEOGEOGRAPHY OF THE LAKE WORTH SUBSEA, OKEECHOBEAN SEA

The Lake Worth Subsea flooded southern Florida in two pulses: the first during the mid-Illinoian warm period and the second during the Sangamonian Interglacial Stage. This last transgression occurred during a time of extremely high sea levels and produced the greatest geomorphological development (Figure 7.1). At that time, the older Immokalee Island highlands, the Hendry Platform, and Miccosukee Island had fused into a large, low-relief island surrounded by a wide zone of mangrove forests, here referred to as **Hendry Island**. This large emergent feature experienced karstic dissolution during the late Pleistocene, which produced the myriad of small sinkholes that are preserved today as the rounded cypress heads in the Big Cypress Swamp (with the trees growing in the thicker soils of the buried sink holes) and Lake Trafford in Collier County. The Recent Immokalee Rise, Southwest Slope, Big Cypress Spur, and Reticulated Coastal Swamps are all remnants of Hendry Island. With the high sea levels during the Sangamonian, Cape Sable was isolated from the southern end of Hendry Island, forming the mangrove-covered **Cape Sable Island**.

After long periods of lacustrine, paludal, and estuarine deposition within the Loxahatchee Trough, this lowest area of the older Okeechobean subseas was now almost completely filled with sediments by Sangamonian times. Instead of a deep lagoon, the eastern Lake Worth Subsea was now a wide, shallow platform-like feature, the **Seminole Lagoon**. To the east, the **Palm Beach Archipelago** and **Tomeu Islands** had become a single large complex of mangrove islands and low sandy cays, separated by transverse tidal channels. Extensive beds of Turtle Grass grew on the sheltered western side of these islands, producing the environments that formed the dense shell beds of the Coffee Mill Hammock Member of the Fort Thompson Formation. Farther south, the old Monroe Reef Tract had enlarged and developed into the **Key Largo Reef System** (named for Key Largo, Monroe County), the predecessor of the Recent Florida Keys. These high-energy zonated reefs produced the coral limestones of the Key Largo Formation, the underpinnings of the northern Florida Keys (High and Low Coral Keys). Oölite that formed in the surf zone (reef crest) of the Key Largo Reefs washed into the lagoon areas behind (west) the coral complexes, fanning out into a wide platform, the **Miami Oölite Banks**. These dune-like deposits, similar to those seen in the Recent Bahamas, extended across Florida Bay and into the Gulf of Mexico, forming the oölite islands of the Recent Florida Keys. The Miami Oölite Banks, cut by tidal channels and covered with Turtle Grass beds and *Schizoporella* ectoproct bryozoans, produced the environments that formed the "coral rock" and "coral ridges" of present-day southern Dade County.

The eastern coast of the Sangamonian Floridian Peninsula, from St. Augustine south to the **Loxahatchee Strait**, was lined by long strands of beaches and transverse dunes. The wave-crushed shells and quartz sand of these beach deposits became cemented in subaerial conditions during the sea level lows of the subsequent Wisconsinan I and II Glacial Stages. This formed the coquina beach rock of the Anastasia Formation that crops out all along Recent eastern Florida. The higher sea level stand of the Sangamonian flooded the Okeechobee Plain area to the greatest extent since the Calabrian. This last marine incursion into the Kissimmee Valley allowed mangrove trees to grow as far north as present-day northern Okeechobee County, forming the **Kissimmee Mangrove Forests**. A reddish, peaty facies of the Coffee Mill Hammock Member of the Fort Thompson Formation was deposited within these mangrove environments. Along Sangamonian western Florida, a narrow set of coastal lagoons, the **Sarasota Lagoon System**, formed behind sand-barrier islands and marked the incipiency of the Recent Gulf Lowlands and Gulf Barrier Islands. The thick and well-developed shell beds of the Fort Thompson near Venice, Sarasota County, were deposited in extensive Turtle Grass beds within the Sarasota Lagoon System.

THE FORT THOMPSON FORMATION, OKEECHOBEE GROUP

History of Discovery

In his publication on stratigraphic sections across the Everglades, Sellards (1919) named a distinct set of mixed lithologies seen along the Caloosahatchee River near LaBelle, Hendry County, the "Fort Thompson beds." These beds, composed of freshwater, estuarine, and shallow marine deposits, were overlaid by a thick layer of "shell marl" composed almost entirely of *Chione* clams. This distinctive unit was named the "Coffee Mill Hammock marl" in the same publication. In their review of Florida geology 10 years later, Cooke and Mossom (1929) elevated the "Fort Thompson beds" to formational status and considered the Coffee Mill Hammock to be the upper member of their new formation. In this book, we concur with Cooke and Mossom and keep the "Fort Thompson beds" and the "Coffee Mill Hammock marl" together as a single formation. With the exception of Brooks (1968) and Lyons (1991), who consider them separate formations, most workers in the Everglades area follow Cooke and Mossom's single formation designation.

In his study of the stratigraphy of the Caloosahatchee River area, DuBar (1958) observed that the lower Fort Thompson beds, below the Coffee Mill Hammock Member, were more lithologically complex than described by earlier workers. These contained a mixture of fresh and brackish water deposits and a distinctive layer of scallops, the "*Chlamys* Bed" (now placed in the genus *Argopecten*). DuBar (1958a) later named this lower section of the Fort Thompson Formation the Okaloacoochee Member, after the Okaloacoochee Branch that empties into the Caloosahatchee River near LaBelle. Subsequent studies, summarized by Lyons (1991: 160), have dated the Okaloacoochee at around 300,000 years B.P., corresponding to the brief mid-Illinoian warm time, and the Coffee Mill Hammock at 150,000 years B.P., corresponding to the Sangamonian stage. Here, we consider the Fort Thompson to be a diachronous unit, with Illinoian and Sangamonian components separated by an unconformity.

Lithologic Description and Areal Extent

The Fort Thompson typically consists of thin, intercalated beds with three main lithologies: freshwater calcarenites ("marls") with *Seminolina* planorbid gastropod bioclasts, densely packed beds of the bivalve *Chione elevata* (previously referred to as *Chione cancellata*, which comprises the majority of the carbonates) in a sparse quartz sand matrix, and muddy sand with beds of bivalves, either the mactrid *Rangia cuneata* (Figure 7.3) or the scallop *Argopecten irradians* subspecies. The scallops are typically stained dark gray or black, whereas the *Rangia* and *Chione* are white or pale tannish-white. Locally, the freshwater beds are completely or partially indurated, forming thin limestone units. In the south, under the Atlantic Coastal Ridge in Dade County (subjacent to the Miami Formation), the lower member of the Fort Thompson is present and is composed of a yellow or pale tan sandy limestone filled with the molds of dissolved mollusk shells. Some of these molds are filled with calcite, producing perfect calcitic pseudomorphs of gastropods and bivalves. This same yellow sandy moldic limestone also frequently contains heavily recrystallized corals, mostly the rose coral *Manicina* and the star coral *Montastrea*. The quartz sands of the *Chione* beds are white, yellow, or pale tan, whereas those mixed with the fine calcarenites of the *Rangia* and *Argopecten* beds are light or dark gray. At Palm Beach Aggregates quarries at Loxahatchee, the Fort Thompson is indurated into a cream-yellow sandy limestone with abundant molluscan bioclasts, mostly *Chione*. In the northern part of the Okeechobee Plain, within the Kissimmee River Valley (particularly the Fort Drum area), the Fort Thompson is composed of a reddish-orange sandy peat filled with the gastropod *Melongena corona* and the bivalve *Pseudomiltha floridana*. This lithology represents deposition within the organic-rich estuarine sand flats of the Kissimmee Mangrove Forests.

As can be seen in Figure 7.2 and Figure 7.5, the combined ranges of the two members of the Fort Thompson Formation cover the entire Everglades area, from Pinellas and Hillsborough

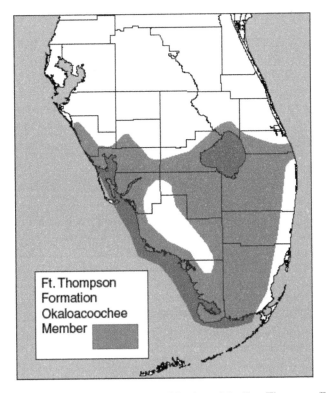

FIGURE 7.2 Areal distribution of the Okaloacoochee Member of the Fort Thompson Formation.

Counties to Monroe County and from northern Highlands and Okeechobee counties to Dade County. The Okaloacoochee Member, deposited during a relatively low sea level stand, is less extensive than the Coffee Mill Hammock Member, which was deposited during the Sangamonian extreme sea level high. The formation, in total, varies in thickness from over 3 m at the Leisey Shell Pit in Hillsborough County (Morgan and Hulbert, 1995) to 5 m at the Venice Shell Pit in Sarasota County, to 2.5 m at the type section in Hendry County, to 1 m along the Miami Canal and at Loxahatchee, Palm Beach County, to 5.5 m at the Six-Mile Bend on old State Road 80 east of Belle Glade, Palm Beach County, to 1.5 m at the Capeletti Brothers Pits west of Miami, Dade County. At most localities, the upper surface of the Coffee Mill Hammock lies between 0.5 and 2.5 m below land surface.

STRATOTYPE

From Cooke (1945: 254): "Old Fort Thompson, the type locality of the formation, lay on the south bank of the Caloosahatchee, 1¹/₂ mile east of the courthouse at LaBelle." This is essentially the same locality as DuBar's station A15 (DuBar, 1958: 231): "SE ¹/₄, SW ¹/₄ Sec. 3, T. 43 S., R. 29 E., Hendry County, Florida, left bank of the Caloosahatchee River approximately 1.5 miles east of the bridge at LaBelle." The top of the formation lies 1 m below land surface.

AGE AND CORRELATIVE UNITS

Based on uranium isotope dating (summarized by Lyons, 1991: 160–161), Fort Thompson deposition extends from the mid-Illinoian Glacial Stage (approximately 300,000 years B.P.) to the end of the Sangamonian Interglacial Stage (75,000 years B.P.). The formation is correlative with the Canepatch and Socastee Formations of the Carolinas.

Fort Thompson Index Fossils

Reflecting the time differential between the two members of the Fort Thompson Formation, the compositions of the marine molluscan faunas of the Okaloacoochee and Coffee Mill Hammock differ greatly. Because of the paucity of large exposures of the Okaloacoochee, its fauna was thought to be essentially the same as the Coffee Mill Hammock by most previous workers. Subsequent research in extensive new Okaloacoochee exposures, such as at Parkland, Broward County (Petuch, 1994, 2004), has shown that the two members have recognizably distinct faunas. The index fossils are listed in the respective section on each member.

The Okaloacoochee Member, Fort Thompson Formation

History of Discovery

As discussed previously, the lower section of the Fort Thompson Formation was first noticed and described from exposures along the Caloosahatchee River in Hendry County (after the river was channelized by the Army Corps of Engineers). DuBar (1958a, 1974) formally named these lower beds, including his "*Chlamys* Bed," the Okaloacoochee Member for the Okaloacoochee Branch, a small tributary of the Caloosahatchee in Hendry County. He also dated the member as being early Sangamonian in age, the overlying Coffee Mill Hammock being late Sangamonian in age. The senior author followed this geochronology, stating that the Okaloacoochee was deposited during the latest Illinoian–earliest Sangamonian (Petuch, 2004: 247). Subsequent biostratigraphic analyses have shown that the member actually is slightly older, dating from the mid-Illinoian.

Lithologic Description and Areal Extent

Condensed here is an extract from DuBar (1974: 224):

> The Okaloacoochee Member is composed of freshwater marl locally divided into upper and lower units by a thin tongue of brackish water marl … The freshwater beds are slightly to moderately indurated; however, locally, the upper few inches to one foot are penetrated by vertical solution holes which commonly are filled with the matrix of overlying units … The clastic fraction of the marls ranges up to 33% by weight, and consists primarily of fine to medium quartz sand. The marl is generally white to light gray or cream in color, but locally it may be yellowish brown or tan … Fossils are abundant and well preserved … In the type area and along Banana Creek, a tongue of brackish water shelly sand four to six inches thick separates the freshwater marl into upper and lower units … The most characteristic species of this unit is the bay scallop *Chlamys irradians concentricus* [authors' note: now known as *Argopecten irradians* unnamed subspecies]. The clastic fraction, which is less than 50% by weight, consists of subangular to subrounded medium quartz sand and minor amounts of silt and clay.

In the northern Everglades, particularly the region south of Lake Okeechobee and the Everglades Agricultural Area (EAA), thin beds of the brackish water bivalve *Rangia cuneata* (Figure 7.3) are found intercalated between freshwater calcarenite layers, indicating rapidly fluctuating salinities within the Seminole Lagoon during Okaloacoochee times.

The member extends from Sarasota County to Monroe County and from southernmost Okeechobee County to Dade County (the yellow sandy moldic limestone found below the Miami Formation). Throughout its range, the Okaloacoochee Member varies in thickness from 2 m at the stratotype in Hendry County to 3 m at the Six-Mile Bend east of Belle Glade, Palm Beach County, to 4 m at Parkland, Broward County, to 1 m along the Miami Canal, western Palm Beach County, to 2 m in the County Pit at Okeechobee, Okeechobee County.

Stratotype

From Dubar (1958: 224): "The member is typically developed on the Caloosahatchee River near Okaloacoochee Branch in Hendry County." The Okaloacoochee Branch is a small stream, tied into

FIGURE 7.3 Detail of a bed of the estuarine mactrid bivalve *Rangia cuneata* in the Okaloacoochee Member of the Fort Thompson Formation. This shell bed was exposed at the Six-Mile Bend east of Belle Glade, Palm Beach County, during the construction of a new canal levee.

a drainage canal system that is fed by flowage from the Okaloacoochee Slough and the marsh system of Collier and Hendry Counties. DuBar's station A14 (DuBar, 1958: 231) contains the best section from the stratotype area: "SW $^1/_4$, SW $^1/_4$ Sec. 2, T. 43 S., R. 29 E., Hendry County, Florida, right bank of the Caloosahatchee River about 150 yards upstream from the point of intersection of a large meander and the canal." The top of the member is 1.5 m below land surface.

Age and Correlative Units

Based on uranium isotope dating, as summarized by Lyons (1991: 160), the Okaloacoochee Member is now known to date from the mid-to-late Illinoian Pleistocene. The member is correlative with the Canepatch Formation of the Carolinas.

Okaloacoochee Index Fossils

The Okaloacoochee contains a molluscan fauna that is transitional between that of the Belle Glade Member of the Bermont Formation and the Coffee Mill Hammock Member of the Fort Thompson Formation. Although it has a high percentage of taxa that are still extant, particularly the bivalves, this member contains a number of endemic taxa and relicts from Bermont time. Of particular interest is the presence of the gastropod genus *Turbinella* and a surviving descendant of the dwarf strombids of the Bermont Formation (the *Strombus lindae* species complex). Both these taxa became extinct in Florida before Sangamonian times and are absent from the Coffee Mill Hammock Member. Both these lineages survive in the Recent Caribbean area, however, with their descendants *Turbinella angulata* ranging from the southern Gulf of Mexico to northern Brazil and *Strombus nicaraguensis* occurring along the Honduran and Nicaraguan coasts. Some of the more abundant and indicative Okaloacoochee species include the following:

Gastropoda
Pyrazisinus gravesae (Figure 7.4C)
Modulus calusa (Figure 7.4F) (still living in Florida Bay Turtle Grass beds)
Strombus alatus kendrewi (Figure 7.4J)
Strombus lindae dowlingorum (Figure 7.4H)
Macrostrombus costatus griffini
Fulguropsis spiratum pahayokee
Sinistrofulgur perversum okeechobeensis (Figure 7.4G)
Melongena (Rexmela) corona subcoronata (Figure 7.4B)
Turbinella wheeleri (Figure 7.4A)
Oliva (Strephona) sayana sarasotaensis (Figure 7.4D) (still living in the Gulf of Mexico off
    western Florida)
Jaspidiconus phluegeri (Figure 7.4I) (still living in the coastal lagoons of Martin and Palm
    Beach Counties)
Bivalvia
Argopecten irradians unnamed subspecies (similar to the living *A. taylorae*)
Euvola ziczac
Rangia cuneata (Figure 7.3)
Mercenaria campechiensis subspecies (Figure 7.4E)

The marine communities of the Okaloacoochee Member were described by the senior author
(Petuch, 2004: 249–251) and include the *Pyrazisinus gravesae* Community (mangrove forests and
intertidal mud flats) and the *Strombus alatus kendrewi* Community (shallow lagoon sand flats and
Turtle Grass beds). These communities were dated as being of "early Sangamonian" age, but should
now be dated as mid-Illinoian. The Okaloacoochee fauna is the first to belong to the Lakeworthian
Mollusk Age.

THE COFFEE MILL HAMMOCK MEMBER, FORT THOMPSON FORMATION

History of Discovery

This uppermost member of the Fort Thompson Formation was described from outcrops near the
Ortona Lock of the Caloosahatchee River, Glades County, by Sellards (1919). Although he considered
it to be a separate formation, most subsequent workers have included it in the Fort Thompson,
as we do.

Lithologic Description and Areal Extent

Cooke (1945: 249) describes it as "a thin shell marl composed chiefly of *Chione cancellata*"
[authors' note: this species is now referred to as *Chione elevata*, with the true *C. cancellata* being
a related but different Caribbean species]. DuBar (1974: 223) gives a more detailed description:

> The fossiliferous facies of the member contains abundant and well-preserved molluscan assemblages
> in a matrix of fine, medium-tan to gray quartz sand. In the type area, the clastic fraction comprises
> about 80% of the unit, and the carbonate fraction is primarily in the form of fossil shells. The most
> common and typical species of the Coffee Mill Hammock are the pelecypod *Chione cancellata* [now
> *Chione elevata*] and *Transenella conradiana* and the gastropods *Bulla occidentalis* and *Olivella mutica*.

We here expand the typical *Chione* bed description of the stratotype to include other lithologies
such as a reddish-orange peaty sand (mangrove peat) with abundant molluscan bioclasts, usually
the bivalve *Pseudomiltha floridana* and the gastropod *Melongena (Rexmela) corona*; a white quartz
sand with abundant molluscan bioclasts, usually *Cerithium muscarum, Pyrazisinus ultimus,* and

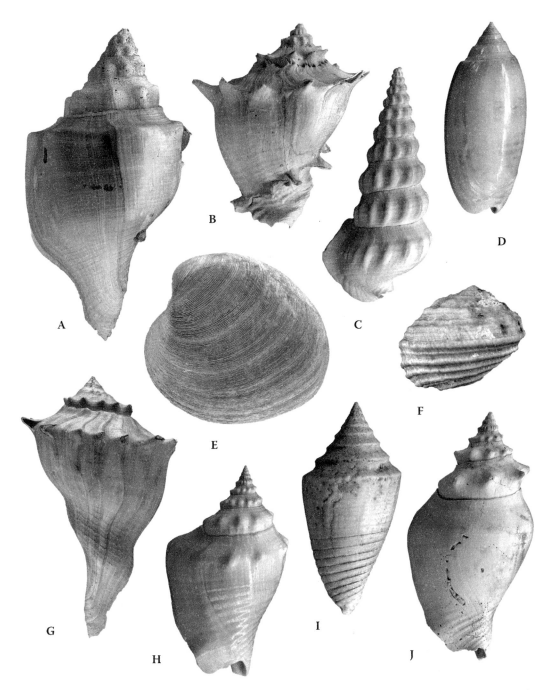

FIGURE 7.4 Index fossils for the Okaloacoochee Member of the Fort Thompson Formation. A = *Turbinella wheeleri* Petuch, 1994, length 258 mm; B = *Melongena (Rexmela) subcoronata* Heilprin, 1886, length 77 mm; C = *Pyrazisinus gravesae* Petuch, 1994, length 55 mm; D = *Oliva (Strephona) sayana sarasotaensis* Petuch and Sargent, 1986, length 41 mm; E = *Mercenaria campechiensis* (Gmelin, 1791) subspecies, length 96 mm; F = *Modulus calusa* Petuch, 1986, width 12 mm; G = *Sinistrofulgur perversum okeechobeensis* Petuch, 1994, length 113 mm; H = *Strombus lindae dowlingorum* Petuch, 2004, length 55 mm; I = *Jaspidiconus phluegeri* Petuch, 2004, length 31 mm; J = *Strombus alatus kendrewi* Petuch, 2004, length 83 mm.

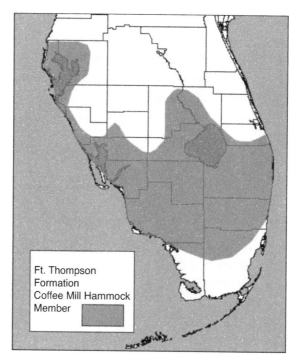

FIGURE 7.5 Areal distribution of the Coffee Mill Hammock Member of the Fort Thompson Formation.

Cerithidea species, occasionally containing petrified (partially silicified) palmetto wood (*Sabal palmetto*; see Petuch, 2004: plate 94); and a pale yellow sandy moldic limestone with partially dissolved mollusks, mostly *Chione elevata*. The reddish peaty facies was deposited within the Sangamonian Kissimmee Mangrove Forests and is the dominant facies in the northern Okeechobee Plain. The white quartz sand–petrified palmetto wood facies was deposited along the eastern side of the Sangamonian Seminole Lagoon, adjacent to the Palm Beach Archipelago.

North of the St. Lucie Inlet of Martin and St. Lucie Counties, the typical Coffee Mill Hammock carbonate bioclast lithology disappears and is replaced by a dark gray sand containing cooler water mollusks and no *Chione*. The molluscan fauna of this unit, which may represent a southern extension of the Socastee Formation of South Carolina, is composed of the gastropods *Busycon carica*, *Busycotypus canaliculatum*, *Neverita duplicata*, and *Oliva (Strephona) sayana* and the bivalves *Mercenaria mercenaria* and *Crassostrea virginica*. This assemblage is still living along the coasts of the Carolinas, Georgia, and northern Florida. The presence of *Busycon carica* along the Sanga-monian coast of St. Lucie and Martin Counties demonstrates that the open oceanic water temperature was slightly cooler than it is today. As in previous times, only the Okeechobean Sea and southwestern Florida remained tropical during the Sangamonian. A similar facies shift is seen along the western coast of Florida around Tampa Bay, where cooler water species, such as the giant muricid gastropod *Hexaplex fulvescens*, make their appearance in Coffee Mill Hammock-equivalent beds. Because both these cooler water faunas and quartz sand beds fall outside the scope of this book, they will not be addressed here. The distribution of the Coffee Mill Hammock Member, as described here, is shown on Figure 7.5.

The Coffee Mill Hammock Member varies in thickness from over 3 m at the Leisey Shell Pit, Hillsborough County, to 3 m in the County Pit, Okeechobee, Okeechobee County, to 1 m in the Palm Beach Aggregates quarries, Loxahatchee, Palm Beach County, to 3 m along State Road 7 (U.S. 441) in Lake Worth, Palm Beach County, to 2.5 m in the Rucks Pit, Fort Drum, Okeechobee County. In Monroe and Dade Counties, the Coffee Mill Hammock grades into the oölitic limestones of the Miami Formation.

Stratotype

The type locality of the Coffee Mill Hammock Member is on the Caloosahatchee River near the Ortona Lock, Glades County. (Dubar's station: A57, NE ¹/₄, SW ¹/₄ Sec. 27, T. 42 S., R. 30 E., Glades County, Florida.) "This is the type locality of the Coffee Mill Hammock marl. The exposures from which a large collection was made occur along the right bank of the Caloosahatchee River about 300 yards downstream from the Atlantic Coast Line Railroad bridge" (Dubar, 1958: 240).

Age and Correlative Units

Based on uranium isotope dating, as summarized by Lyons (1991: 150–161), the Coffee Mill Hammock Member is now known to date from the Sangamonian Pleistocene. The member is correlative with the Socastee Formation of the Carolinas.

Coffee Mill Hammock Index Fossils

This last geologic unit of the Lakeworthian Mollusk Age contains a distinctive fauna composed of three components: species still living in southern Florida, living Caribbean species now extinct in Florida, and Coffee Mill Hammock endemics. Some of the more abundant and indicative species include the following:

Species extant in Recent Florida
Gastropoda
Cerithidea costata (Figure 7.7E)
Batillaria minima (Figure 7.7H)
Cerithium muscarum (Figure 7.7F)
Cerithium algicola
Cerithium floridanum
Modulus pacei (Figure 7.7J) (Recent southeastern Florida only)
Strombus alatus (Figure 7.6I)
Strombus (Lobatus) raninus
Eustrombus gigas (Figure 7.7B)
Macrostrombus costatus
Macrocypraea cervus
Cassis spinella
Eupleura sulcidentata
Vokesinotus perrugatus (Figure 7.6D) (Recent western Florida only)
Fasciolaria tulipa
Fasciolaria (Cinctura) hunteria
Triplofusus giganteus
Fulguropsis spiratum
Sinistrofulgur sinistrum (Figure 7.6F)
Melongena (Rexmela) corona (Figure 7.6H)
Melongena (Rexmela) corona winnerae (Figure 7.6G) (lagoons and tidal creeks of Recent
    Martin and Palm Beach Counties only)
Phrontis vibex
Solenosteira cancellarius (Figure 7.6E) (Recent western Florida only)
Columbella rusticoides (Figure 7.7G)
Vasum muricatum
Oliva (Strephona) sayana
Olivella mutica
Gradiconus patglicksteinae (Figure 7.7A) (Recent eastern Florida only)

FIGURE 7.6 Index fossils for the Coffee Mill Hammock Member of the Fort Thompson Formation. A = *Melongena melongena* (Linnaeus, 1758), length 109 mm; B = *Cerithidea beattyi lakeworthensis* Petuch, 2004, length 11 mm; C = *Pyrazisinus ultimus* Petuch, 2004, length 53 mm; D = *Vokesinotus perrugatus* (Conrad, 1836), length 22 mm; E = *Solenosteira cancellarius* (Conrad, 1846), length 29 mm; F = *Sinistrofulgur sinistrum* (Hollister, 1958), length 212 mm (this species, which is still living along the southeastern U.S. and the Gulf of Mexico, is often incorrectly referred to in the literature as *S. contrarium*, which is a Pliocene species found in the Yorktown and Duplin Formations and the Pinecrest Member of the Tamiami Formation); G = *Melongena*

FIGURE 7.6 (continued) *(Rexmela) corona winnerae* Petuch, 2004, length 150 mm; H = *Melongena (Rexmela) corona* (Gmelin, 1791), length 82 mm; I = *Strombus alatus* Gmelin, 1791, length 80 mm; J = *Chione elevata* (Say, 1822), length 23 mm (this species has incorrectly been referred to the taxon *Chione cancellata*, which is a related species found in the Caribbean Sea).

Jaspidiconus stearnsi (Recent western Florida only)
Spuriconus spurius
Melampus coffeus (Figure 7.7D)
Melampus monilis (Figure 7.7C)
Bulla occidentalis
Bivalvia
Pseudomiltha floridana (Recent western Florida to Texas only)
Anomalocardia aubreyana
Chione elevata (Figure 7.6J)
Mercenaria campechiensis

Species living in the Recent southern Gulf of Mexico and Caribbean
Gastropoda
Melongena melongena (Figure 7.6A)
Cerithidea beattyi lakeworthensis (Figure 7.6B) (extinct Floridian subspecies of the Caribbean
　C. beattyi)

Species restricted to the Coffee Mill Hammock Member (endemics)
Gastropoda
Cerithidea scalariformis palmbeachensis (Figure 7.7I)
Pyrazisinus ultimus (Figure 7.6C)
Bivalvia
Anomalocardia hendryana

The marine communities of the Coffee Mill Hammock Member were described by the senior author (Petuch, 2004: 250–253) and include the *Pyrazisinus ultimus* Community (mangrove forests and intertidal mud flats) and the *Chione elevata* Community (shallow lagoon and Turtle Grass beds).

THE MIAMI FORMATION, OKEECHOBEE GROUP

History of Discovery

The first geologist to study the limestones of the southern Atlantic Coastal Ridge and the Florida Keys was Samuel Sanford of the Florida Geological Survey. In 1909, he named the surficial oölitic limestone of the Atlantic Coastal Ridge the "Miami oölite" and that of the central and southern Keys the "Key West oölite". He separated the two formations by differing amounts of quartz sand, with the Miami having a higher percentage than the Key West. Recognizing that the two limestones were of the same age and that they formed a lithologic continuum, Cooke and Mossom (1929) later combined the Key West with the Miami, considering it to be simply a facies.

Studying analogue environments in the Bahamas, Hoffmeister et al. (1967) later recognized two broad lithofacies and biofacies within the Miami Formation: an oölitic facies composed of unconsolidated, partially consolidated (by calcite microspar cementing), or indurated (oömoldic) limestones (Figure 7.9) and a "bryozoan facies" made up primarily of interlocking branches of the cheilostome ectoproct ("bryozoan") *Schizoporella floridana*. The oölitic "dunes" of the Miami area (particularly near Coconut Grove) are frequently cross-bedded (Figure 7.10), with dips of up to 30°. In southern and southwestern Dade County (particularly in the Homestead area), the

FIGURE 7.7 Index fossils for the Coffee Mill Hammock Member of the Fort Thompson Formation. A = *Gradiconus patglicksteinae* (Petuch, 1987), length 38 mm (may be a subspecies of *G. anabathrum*); B = *Eustrombus gigas* (Linnaeus, 1758), length 178 mm; C = *Melampus monilis* (Bruguiere, 1789), length 10 mm; D = *Melampus coffeus* (Linnaeus, 1758), length 12 mm; E = *Cerithidea costata* (daCosta, 1778), length 16 mm; F = *Cerithium muscarum* Say, 1832, length 24 mm; G = *Columbella rusticoides* (Heilprin, 1886), length 15 mm; H = *Batillaria minima* (Gmelin, 1791), length 16 mm; I = *Cerithidea scalariformis palmbeachensis* Petuch, 2004, length 23 mm; J = *Modulus pacei* Petuch, 1987, length 16 mm.

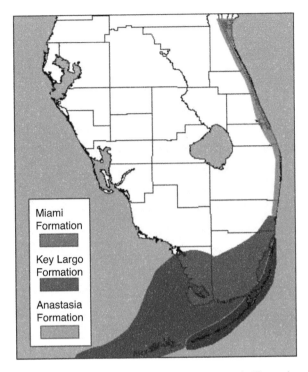

FIGURE 7.8 Areal distributions of the Miami, Key Largo, and Anastasia Formations.

bryozoan facies is often referred to locally as "coral rock" (the major constituent of the Coral Castle tourist attraction near Homestead). This is not to be confused with the true coral rock of the Key Largo Formation.

LITHOLOGIC DESCRIPTION AND AREAL EXTENT

As discussed previously under the description of the Lake Worth Subsea, the Miami Formation is the result of deposition of oöids produced on, and behind, the Key Largo Reef System during Sangamonian times. These accumulated in immense dune-like structures, similar to those on the Recent Great Bahama Bank. The best lithologic description of the Miami Formation, summarizing previous works, was given by DuBar (1974: 228). This is repeated here in a condensed form:

> The fresh rock is relatively soft, but hardens upon exposure to the atmosphere. The chief constituent particles are oöids, pellets, and skeletal sand ... Many of the originally aragonitic oöids and pellets have been replaced by calcite. With depth they are increasingly embedded in a crystalline calcite matrix. Below the water table, both oöids and pellets have completely dissolved away, leaving only spherical and ellipsoidal cavities so that the rock becomes oömoldic. The calcitic matrix is a post-depositional precipitation filling the empty interstices ... In the Atlantic Coastal Ridge, the bryozoan facies underlies the oölitic facies ... the facies consists predominantly of zoaria of the cheilostome bryozoan *Schizoporella floridana* (up to 80% by volume). The bryozoans are mixed with oöids, pellets, skeletal sand, polychaete worm tubes, and a sparse molluscan assemblage ... The hard calcitic matrix forms a cellular or vesicular structure.

As can be seen in Figure 7.8, the Miami Formation extends from extreme southeastern Broward County to most of Monroe County. In Monroe County, the Miami underlies Cape Sable, all of Florida Bay, and the Oölite Keys (Big Pine Key area to west of Key West). Within this range, the

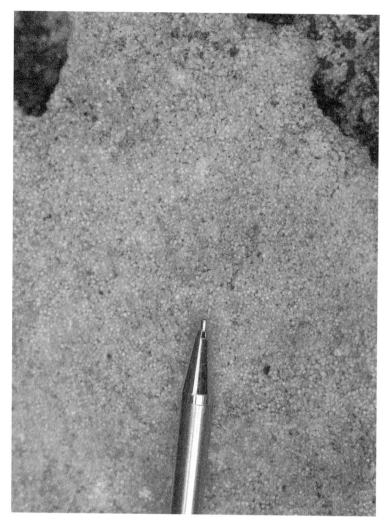

FIGURE 7.9 Detail of a fresh exposure of the Miami Formation, showing the well-defined individual oöids. Photo by Anton Oleinik.

bryozoan facies makes up most of the lithofacies, covering an area of over 3000 km². The Miami Formation averages 2–5 m thickness throughout most of its areal distribution.

STRATOTYPE

The type locality of the Miami Formation is generally considered to be the exposure along Silver Bluff, on Bayshore Drive in Coconut Grove, Miami, Dade County.

AGE AND CORRELATIVE UNITS

As summarized by DuBar (1974: 229), uranium isotope dating of the Miami Formation shows that it ranges in age from 120,000–130,000 years B.P., placing it in the Sangamonian Interglacial Stage of the late Pleistocene. The formation is correlative with the Socastee Formation of the Carolinas, the Coffee Mill Hammock Member of the Fort Thompson Formation of southern Florida, the Anastasia Formation of eastern Florida, and the upper beds of the Key Largo Formation of southern Florida.

FIGURE 7.10 View of an exposure of the Miami Formation in Coral Gables, Miami, Dade County, showing typical cross-bedding seen in the oölitic "dunes." Photo by Anton Oleinik.

Miami Index Fossils

Although branching colonies of the cheilostome ectoproct *Schizoporella floridana* make up most fossils preserved in the Miami Formation, rare specimens of gastropod and bivalve mollusks do occur. These are primarily the large queen conch *Eustrombus gigas* (Figure 7.7B) and the bivalve *Codakia orbicularis*. Both species are common in Recent Caribbean and southern Floridian Turtle Grass beds. DuBar (1974: 228) comments that "two common growth forms are exhibited by the bryozoan colonies. The forms on the Atlantic Coastal Ridge area are rough, irregular masses up to 4 feet in length. The zoaria of the western shelf area are also irregular, but usually are no more than 5 inches in length. It is probable that the zoaria of both groups formerly encrusted clumps of *Thalassia*."

THE ANASTASIA FORMATION, OKEECHOBEE GROUP

History of Discovery

Elias Sellards, in his early pioneer studies, was the first to recognize and name "the extensive deposit of coquina rock found along the east coast." With the best outcrops of this coquina rock

FIGURE 7.11 View of outcrops of the Anastasia Formation at Jupiter Blowing Rocks, Palm Beach County. This sandy coquina constitutes the "Palm Beach Limestone" facies of the Anastasia Formation. Photo by Anton Oleinik.

being found on Anastasia Island, opposite St. Augustine, Sellards named this unit the Anastasia Formation. Around the same time, Sanford (1913) named an indurated sandy calcarenite from the Palm Beach County coastline, calling it the "Palm Beach Limestone" (Figure 7.11 and Figure 7.12). Later, Cooke and Mossom (1929) considered this sandier southern component to be simply a local facies and included Sanford's Palm Beach Limestone in Sellard's Anastasia Formation. In this book, we concur with Cooke and Mossom and recognize a single narrow formation, running down the coast from St. John's County to northernmost Broward County.

LITHOLOGIC DESCRIPTION AND AREAL EXTENT

Throughout its range, the Anastasia Formation is relatively homogeneous, being composed of rounded molluscan shell fragments and quartz sand cemented together by calcite spar and iron oxide. Typically, the Anastasia consists of one or more beds of coquina, with shell fragments varying from large (2–5 cm) to small (0.5–1 cm). Often, large beach-worn gastropods are embedded in the matrix of small rounded shell fragments and flakes. Quartz sand amounts are variable, with some facies (such as at the stratotype) being composed almost entirely of cemented large fragments and unbroken shells and with other facies being almost a quartz sandstone, with the shell fragments reduced to small rounded grains. Along Palm Beach County, the quartz sand fraction is high, producing a dense sandy calcarenite with lesser amounts of finely crushed shell fragments. This facies of the Anastasia weathers quickly, being covered with microkarstic solution holes. Along Jupiter Island, Palm Beach County (particularly Jupiter Blowing Rocks, Figure 7.11 and Figure 7.12),

FIGURE 7.12 View of a large wave-eroded "blow hole" in the Anastasia Formation outcrops at Jupiter Blowing Rocks, Palm Beach County. Photo by Anton Oleinik.

the Anastasia forms a 3 m-high cliff riddled with wave "blow holes." Unweathered sections of the Anastasia are variable in color, with shell fragments ranging from golden orange to tan, to light grayish-brown, to gray. Often the shell fragments retain their original color, adding flecks of red, white, or black to the base color of the coquina.

The Anastasia crops out all along the eastern coast of Florida, but rarely more than 5 km west of the Intracoastal Waterway (Cooke, 1945: 266) (Figure 7.8). The formation extends from St. John's County to northernmost Broward County. At some localities, such as at Bathtub Beach near Stuart, Martin County and at Jupiter Blowing Rocks, Palm Beach County, the Anastasia forms low cliffs and extensive raised beach rock terraces and pavements. South of this area to Hillsboro Inlet, the formation crops out as sporadic, small, low cliff-like features (such as at Gulf Stream and Red Reef Parks, Palm Beach County). Throughout its range, the Anastasia varies in thickness from 10 m in St. John's County, to 1 m along Flagler County, to 3–4 m at Jupiter Blowing Rocks, Palm Beach County, to over 7 m at the Palm Beach road cut, Palm Beach Island, Palm Beach County. The Anastasia Formation underlies Singer Island, Palm Beach Island, and the islands of Lake Worth in Palm Beach County.

STRATOTYPE

The type locality is the stretch of coquina outcrops near the lighthouse on Anastasia Island, St. John's County, Florida.

AGE AND CORRELATIVE UNITS

Based on its interfingering with the uranium-dated Miami Formation and Coffee Mill Hammock Member of the Fort Thompson Formation, the Anastasia Formation is now known to be of Sanga-monian Pleistocene age. The beach sands and shell fragments were deposited at this time, but were later cemented during the sea level drop caused by the Wisconsinan glaciations. The Anastasia is correlative with the Socastee Formation of the Carolinas, the Key Largo Formation of southern Florida, and the interfingering units mentioned previously.

ANASTASIA INDEX FOSSILS

Because the formation is composed primarily of broken, wave-rounded shell fragments, few iden-tifiable index fossils are encountered. At localities along Indian River, St. Lucie, and Martin Counties, large, fairly intact specimens of the busyconid gastropod *Sinistrofulgur sinistrum* (Figure 7.6F) have been found embedded in the shell fragment matrix.

THE KEY LARGO FORMATION, OKEECHOBEE GROUP

HISTORY OF DISCOVERY

In his survey of the topography and surficial geology of southern Florida, Sanford (1909) first described the coral limestone found in the near surface of the northern and middle Florida Keys. This was named the "Key Largo limestone" for Key Largo, the longest of the Florida Keys. As the geologist assigned to the construction of the Florida East Coast Railway, Sanford had unprecedented opportunities to study sections of the Key Largo Formation that were exposed in canal walls and manmade cuts between keys. In their classic study of the geology and origins of the Florida Keys, Hoffmeister and Multer (1968) were the first to describe the detailed lithologic and paleoecological features of the Key Largo as well as its thickness and structure.

LITHOLOGIC DESCRIPTION AND AREAL EXTENT

The following is a condensed extract from the lithologic summary of DuBar (1974: 230):

> The limestone is composed of an *in situ* organic framework of hermatypic corals and trapped interstitial calcarenite … cementing and binding organisms, such as crustose coralline algae and milleporid corals, form only a small portion of the reef … the coral heads are surrounded by poorly sorted calcarenite and that well-sorted calcarenite predominates in channels … aragonite and high magnesium calcite probably were the chief original mineralogic components. The original high-magnesium fossils have been altered in place to low-magnesium calcite through the removal of excess magnesium within the meteoric zone. Aragonite is being dissolved currently by meteoric water, and the dissolved calcium carbonate is being redeposited near the site of dissolution as low-magnesium calcite cement. Many originally aragonitic fossils now occur as molds of sparry calcite … five hermatypic corals account for 30% of the rock volume … *Montastrea annularis* is by far the most important coral species.

The coral limestone varies in color from white, to pale yellow, to light gray.

As shown in Figure 7.8, the Key Largo Formation underlies only the High and Low Coral Keys of the southeasternmost Dade and Monroe Counties, from Key Biscayne and Soldier Key to Bahia Honda Key. Throughout this range, the fossil reef tract is exposed above water in a narrow strip of 5 km width. The Key Largo extends at least another 8 km seaward, to the edge of the living Recent reef tract. On the Florida Bay side, the Key Largo interfingers with the oölites of the Miami Formation. The formation ranges in thickness from 20–50 m, with the thickest limestones being in the western part. The older, lower calcarenite and calcilutite facies, occurring 60–70 m below Key Largo and Key West (Hoffmeister and Multer, 1968), were part of the Monroe Reef Tract and date from the mid-Pleistocene. These represent a reefal facies of the Bermont Formation (see Chapter 6).

STRATOTYPE

The type section for the Key Largo Formation is in the Windley Key Quarry, now a geologic preserve, on Windley Key, between Plantation Key and Upper Matecumbe Key.

AGE AND CORRELATIVE UNITS

Based on uranium isotope dating (Osmond et al., 1965), the upper part of the Key Largo Formation is now known to have been deposited 120,000 years B.P., placing it in the Sangamonian Pleistocene. This upper section is correlative with the Miami Formation, the Anastasia Formation, and the Coffee Mill Hammock Member of the Fort Thompson Formation, all of southern Florida, and the Socastee Formation of the Carolinas. The lower section of the Key Largo Formation dates from the mid-Illinoian Pleistocene and is correlative with the Okaloacoochee Member of the Fort Thompson Formation of southern Florida and the Canepatch Formation of the Carolinas.

KEY LARGO INDEX FOSSILS

Only five species of corals, all massive forms, dominate the Key Largo Formation. These include the following:

Cnidaria–Scleractinia
Diploria clivosa
Diploria labyrinthiformis
Diploria strigosa
Montastrea annularis
Porites astreoides

All these are reef platform and back-reef species, indicating that the reef crest existed farther seaward, under the present-day Hawk Channel. Hoffmeister and Multer (1968) discovered the reef crest species *Acropora palmata* in cores taken from the outer edge of the platform, demonstrating the existence of this reef zone. The Key Largo reef crest, however, was largely removed by wave action and dissolution during the Wisconsinan sea level lows, leaving the reef platform and back reef as the only remnants of a once-wider feature. At the Windley Key Quarry, the encrusting bivalves *Chama macerophylla* and *Spondylus americanus* can be seen to have grown on the eroded undersides of some of the massive coral heads.

THE SURFICIAL COMPLEX

The upper 1–5 m of the Everglades region is covered by three main types of sediments: quartz sand, fresh or brackish water lime muds (calcilutites), and peats and mucks consisting of decomposed plant matter. These three sediment types and their mixtures make up the Surficial Complex. In this book, we concentrate only on the surficial sediments of the Everglades Basin and around Lake Okeechobee, and not those found farther west or north. All these were deposited during the latest Pleistocene (latest Wisconsinan II Glacial Stage) and the Holocene, and some are being

FIGURE 7.13 View of an exposure of the Lake Flirt Formation in a canal near the Okeelanta sugar mill, Palm Beach County. Here, the three thin layers of the Lake Flirt can be seen below, and in contact with, the black surface layer of Everglades peat. These indurated beds form a prominent shelf that is unconformably underlain by estuarine deposits of the Fort Thompson Formation.

deposited in the present. Soils such as the Everglades peat, and other peats and mucks, will not be covered in this book.

THE LAKE FLIRT FORMATION, SURFICIAL COMPLEX

HISTORY OF DISCOVERY

In 1919, Elias Sellards described a distinctive freshwater calcarenite found in Lake Flirt, a shallow expansion of the Caloosahatchee River between Lake Hicpochee, Glades County, and LaBelle, Hendry County. Named the "Lake Flirt marl," the formation was later recognized (Cooke, 1945: 312) as extending throughout the Everglades, underlying the Everglades peat (Figure 7.13) and the mucks surrounding Lake Okeechobee. Lake Flirt was later drained during the channelization of the Caloosahatchee River, and its remnants are now bisected by State Highway 80 east of LaBelle.

LITHOLOGIC DESCRIPTION AND AREAL EXTENT

Condensed here is a text from the lithologic summary of DuBar (1974: 231): "The formation consists of thin beds of mucky dark sands and marly shell beds which characteristically contain abundant freshwater gastropods and a few freshwater vertebrate fossils ... in the type area, three Lake Flirt marl beds are separated by thin mucky units." In some localities, these "marls" are made up almost entirely of *Chara* charaphyte phytoliths. In the EAA, the soft calcarenites are indurated into a friable-to-dense freshwater limestone (Figure 7.13). This indurated facies is white to cream-white in color and is typically composed of three thin, distinct layers. The upper surface of the Lake Flirt is often heavily indurated, forming a thick duricrust that is pitted with microkarstic solution holes. These thick duricrusts are often stained yellow or pale orange by limonite and are referred to locally as "cap rock."

The Lake Flirt Formation extends from the area near LaBelle, Hendry County, southward to Monroe (the "Flamingo marl" facies) and northwestern Dade Counties. Throughout its range, the Lake Flirt varies in thickness from 1 to 3 m. Calcarenites of the Lake Flirt type are still being deposited in Lake Hicpochee, Glades County, and within the charaphyte-dominated pools of the Everglades.

STRATOTYPE

The type section of the Lake Flirt Formation is here designated as DuBar's station A59 (DuBar, 1958: 240) on the Caloosahatchee River near the Ortona Lock, Glades County "NE $\frac{1}{4}$, SW $\frac{1}{4}$ Sec. 28, T. 42 S., R. 30 E., Glades County, Florida. Section on left bank of the Caloosahatchee River. Only the Lake Flirt marl beds are exposed."

AGE AND CORRELATIVE UNITS

Based on Carbon 14 dating, as summarized by DuBar (1974: 231), the Lake Flirt Formation was deposited between 20,900 years and 3,800 years B.P., spanning the time from the late Wisconsinan II Glacial Stage of the Pleistocene to the mid-Holocene.

LAKE FLIRT INDEX FOSSILS

The Lake Flirt was deposited within a series of small lakes scattered around the Everglades during the latest Pleistocene and Holocene times. These small lakes, together with the larger Lake Flirt, contained a molluscan fauna that is extant in the Recent Everglades. These Lake Flirt and Recent species are all descendants of the faunas of Lakes Okeelanta and Pahayokee. The Lake Flirt and the living species of planorbid gastropods are described by Pilsbry (1934). The more abundant Lake Flirt species include the following:

Gastropoda
Planorbella duryi
Planorbella eudiscus
Planorbella intercalare
Planorbella normale
Seminolina scalare
Seminolina seminole
Pomacea paludosa
Viviparus georgianus

THE PAMLICO FORMATION, SURFICIAL COMPLEX

HISTORY OF DISCOVERY

The name "Pamlico Formation" has been applied to the surficial sands and beach dunes of southern Florida by almost all previous workers. As discussed by Cooke (1945: 297), the formation was originally proposed for deposits of late Pleistocene quartz sand, fine sandy loam, clay, and gravel found in the vicinity of Pamlico Sound, North Carolina.

LITHOLOGIC DESCRIPTION AND AREAL EXTENT

Throughout the late Pleistocene, the quartz sand component of the Pamlico Formation was carried down the southeastern U.S. coast by longshore transport and deposited as beach sand along eastern

FIGURE 7.14 View of an exposure of the Florida Pamlico Formation in the Palm Beach Aggregates, Inc., pit #6, Loxahatchee, Palm Beach County, showing intercalated beds of quartz sand and clays. Photo by Anton Oleinik.

Florida. Much of this sand was later (latest Pleistocene–Holocene) blown inland to form coastal transverse dunes and thin surficial sand layers on top of older features. These late Quaternary sands were later mixed with the older sands that had covered the buried paleoislands of the southern Atlantic Coastal Ridge. Because of the mixed origins and ages of the Everglades surficial sands, we feel that the name Pamlico Formation should be applied to them only in the broadest and most informal sense.

The Florida Pamlico is composed almost entirely of well-rounded quartz sand, fine to medium, varying in color from white to yellow, tan, brown, and gray. In some localities, the Pamlico contains lenses of black sand derived from ferromagnesian and titanium minerals. In the higher elevation areas of the Atlantic Coastal Ridge, the Pamlico sands often contain intercalated clay and lime mud units (Figure 7.14). These represent local deposition within early Holocene ponds and marshlands. In other areas, such as in the Caloosahatchee Valley (DuBar, 1974: 236) and Okeechobee Plain, the Pamlico has mixed with decomposed vegetation to produce a dark brown or black carbonaceous sand.

The Florida Pamlico extends over the entire area covered by this book, where it ranges in thickness from 1 m along the Caloosahatchee River to 2 m in the Rucks Pit, Fort Drum, Okeechobee County, to over 3 m in the Palm Beach Aggregates quarries, Loxahatchee, Palm Beach County. The large transverse beach dunes of Palm Beach County, which extend from Juno Beach to West Palm Beach, may be partially, or entirely, made up of the Florida Pamlico Formation. Within the Everglades proper, the Pamlico sands are replaced by soils such as the Everglades Peat, Loxahatchee Peat, Okeelanta Peaty Muck, and the Okeechobee Pond Apple Muck. All these soils interfinger with the Pamlico sands on the western edge of the Atlantic Coastal Ridge and along Lake Okeechobee.

AGE AND CORRELATIVE UNITS

The Florida Pamlico dates from the latest Wisconsinan II Glacial Stage of the Pleistocene to the Holocene. The Floridian surficial sands are, at least in part, correlative with the Pamlico Formation of North Carolina.

PAMLICO INDEX FOSSILS

The Florida Pamlico sands contain a rich fossil vertebrate fauna, including mammals of the latest Rancholabrean age, and chelonian and crocodilian reptiles. On the eastern side of the Everglades, along the Atlantic Coastal Ridge of Palm Beach and Broward Counties, frequently encountered vertebrates include mastodon and mammoth prodoscideans (*Mammut* and *Mammuthus*; mostly teeth, long bones, vertebrae, and tusk fragments), horses (*Equus*; mostly teeth and leg bones), ground sloths (*Megalonyx*; mostly teeth and claw cores), alligators (*Alligator*; mostly teeth and osteoderms), and giant tortoises (*Hesperotestudo*; carapace and plastron fragments). These are often found in construction excavations and shallow canal digs. Sand-encased fossil crabs (*Oxypode*) have also been found in some Pamlico exposures, particularly along Vero Beach (see Petuch, 2004: plate 91).

The distributions and thicknesses of the Pamlico sands and the Everglades Peat correspond to the near subsurface topography set down by the underlying buried pseudoatoll. As shown in Figure 7.15, the high structural relief established in the Tamiami and Caloosahatchee Subseas was greatly reduced by infilling during the times of the Loxahatchee, Belle Glade, and Lake Worth Subseas. In the Recent, the sheet flow of the Everglades still conforms to the muted depression of the Plio–Pleistocene Loxahatchee Trough and Seminole Lagoon. The great urban expanses of West Palm Beach, Fort Lauderdale, Miami, Naples, and Fort Myers, all conform to the high areas formed by the pseudoatoll reef tracts of the Tamiami Subsea. The DeSoto and Immokalee Deltas of the Charlotte and Murdock Subseas and the Hendry Platform of the Tamiami and Caloosahatchee Subseas, altogether, produced the higher flatlands of the EAA, ultimately providing the soils that support the sugar industry in Palm Beach, Hendry, and Glades Counties. The legacy of the Okeechobean Sea's ancient geomorphology is manifested in all the aspects of today's Everglades.

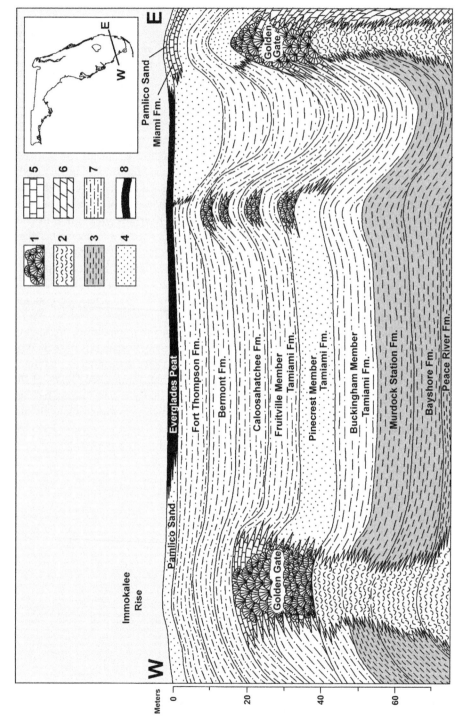

FIGURE 7.15 Generalized cross section across the Everglades, from Naples, Collier County (W) to Fort Lauderdale, Broward County (E), showing the stratigraphy and structures of the upper 60 m. Lithologies are as follows: 1 = coral reefs and bioherms; 2 = oyster beds; 3 = phosphatic clays; 4 = quartz sand with shells; 5 = limestones; 6 = dolomitized limestones; 7 = calcilutites; quartz sand, and shells; 8 = peats and muck. Vertical scale greatly exaggerated. Drafting by Anton Oleinik.

References

Applegate, A.V., April 28, 1986, Corkscrew field boosts South Florida drilling hopes, *Oil and Gas Journal,* 104–106.

Bottomley, R., Grieve, R.A.F., York, D., and Masaitis, V., 1997, The age of the Popigai impact event and its relationship to events at the Eocene-Oligocene boundary, *Nature,* 388: 365–368.

Brooks, H.K., 1968, The Plio-Pleistocene of Florida, with special reference to the strata outcropping on the Caloosahatchee River, in Perkins, R.D., Ed., *Late Cenozoic Stratigraphy of Southern Florida — A Reappraisal, with Notes on the Sunoco-Felda and Sunniland Oil Fields,* pp. 3–42, Guidebook, Second Annual Field Trip, February 1968, Miami Geological Society, Coral Gables, FL.

Brooks, H.K., 1974, Lake Okeechobee, in Gleason, P.J., Ed., *Environments of South Florida: Present and Past,* Miami Geological Society Memoir 2, Miami Geological Society, Coral Gables, FL, pp. 256–286.

Campbell, L.D., 1993, *Pliocene Molluscs from the Yorktown and Chowan River Formations of Virginia,* Virginia Division of Mineral Resources, Publication 127, Charlottesville, VA, 259 pp.

Cooke, C.W., 1945, *Geology of Florida,* Geological Bulletin 29, Florida Geological Survey, State of Florida Department of Conservation, Tallahassee, FL, 339 pp.

Cooke, C.W. and Mansfield, W.C., 1936, Suwannee Limestone of Florida (abstract), Geological Society of America Proceedings for 1935, pp. 71–72.

Cooke, C.W. and Mossom, S., 1929, *Geology of Florida,* Florida Geological Survey Annual Report 20, Florida Geological Survey, Tallahassee, FL, pp. 29–227.

Cunningham, K.J., Locker, S.D., Hine, A.C., Bukry, D., Barron, J.A., and Guertin, L.A., 2003, Interplay of Late Cenozoic Siliciclastic and Carbonate response on the Southeast Florida Platform, *Journal of Sedimentary Research,* 73(1): 31–46.

Dall, W.H., 1887, Notes on the Geology of Florida, *American Journal of Science,* (Series 3), 34: 161–170.

Dall, W.H., 1890, Contributions to the Tertiary Fauna of Florida, with especial reference to the Miocene Silex Beds of Tampa and the Pliocene Beds of the Caloosahatchie River. Part 1, *Transactions of the Wagner Free Institute of Science of Philadelphia,* 3(1): 1–200.

Dall, W.H., 1892, Contributions to the Tertiary Fauna of Florida, with especial reference to the Miocene Silex Beds of Tampa and the Pliocene Beds of the Caloosahatchie River. Part 2, *Transactions of the Wagner Free Institute of Science of Philadelphia,* 3(2): 201–474.

Dall, W.H., 1903, Contributions to the Tertiary Fauna of Florida, with especial reference to the Miocene Silex Beds of Tampa and the Pliocene Beds of the Caloosahatchie River. Part 6, Concluding the Work, *Transactions of the Wagner Free Institute of Science of Philadelphia,* 3(6): 1219–1654.

Dall, W.H., 1915, A monograph on the molluscan fauna of the *Orthaulax pugnax* Zone of the Oligocene of Tampa, Florida, *U.S. National Museum Bulletin,* 90: 1–173.

Douglas, M.S., 1947, *The Everglades: River of Grass* (revised ed.; 1992), Mockingbird Books, Marietta, GA, 308 pp.

Dowsett, H.J. and Cronin, T.M., 1990, High eustatic sea level during the middle Pliocene: evidence from the southeastern U.S. Atlantic Coastal Plain, *Geology,* 18(5): 435–438.

DuBar, J.R., 1958, *Stratigraphy and Paleontology of the Late Neogene Strata of the Caloosahatchee River Area of Southern Florida,* Geological Bulletin 40, Florida Geological Survey, Florida State Board of Conservation, Tallahassee, FL, 267 pp.

DuBar, J.R., 1958a, Neogene stratigraphy of southwestern Florida, *Transactions of the Gulf Coast Association of Geological Societies,* 8: 129–155.

DuBar, J.R., 1974, Summary of the Neogene Statigraphy of southern Florida, in Oaks, R.Q. and DuBar, J.R., Eds., *Post-Miocene Stratigraphy, Central and Southern Atlantic Coastal Plain,* Utah State University Press, Provo, UT, pp. 206–231.

Enos, P. and Perkins, R.D., 1977, *Quaternary Sedimentation in South Florida,* Geological Society of America Memoir 147, 198 pp.

Ganapathy, R., 1982, Evidence for a Major Meteorite Impact on the Earth 34 Million Years Ago: Implication for Eocene Extinctions, *Science,* 216, pp. 885–886.

Glass, B.P. and Zwart, M.J., 1979, North American microtektites in deep sea drilling project cores from the Caribbean Sea and Gulf of Mexico, *Geological Society of America Bulletin* (Part 1), 90: 595–602.

Gleason, P.J., Ed., 1984, *Environments of South Florida: Present and Past II*, Miami Geological Society, Coral Gables, FL, 551 pp.

Haq, B.U., Hardenbol, J., and Vail, P.R., 1988, Mesozoic and Cenozoic chronostratigraphy and cycles of sea-level change, in Wilgus, C.K., Ed., *Sea-level Changes — An Integrated Approach*, Society of Economic Paleontologists and Mineralogists Special Report, 42: 71–108.

Heilprin, A., 1886, Explorations on the West Coast of Florida and in the Okeechobee Wilderness, *Transactions of the Wagner Free Institute of Science of Philadelphia*, 1: 366–506.

Hoerle, S.E., 1970, Mollusca of the "Glades Unit" of southern Florida. Part 2: List of the Molluscan Species from the Belle Glade Rock Pit, Palm Beach County, FL, *Tulane Studies in Geology and Paleontology*, 8(1,2): 56–68.

Hoffmeister, J.E. and Multer, H.G., 1968, Geology and origin of the Florida Keys, *Geological Society of America Bulletin*, 79: 1487–1502.

Hoffmeister, J.E., Stockman, K.W., and Multer, H.G., 1967, Miami limestone of Florida and its Recent Bahamian counterpart, *Geological Society of America Bulletin*, 78: 175–190.

Hollister, S.C., 1971, New *Vasum* species of the subgenus *Hystrivasum*, *Bulletins of American Paleontology*, 58(262): 209–304.

Hunter, M.E., 1968, Molluscan Guide Fossils in Late Miocene Sediments of Southern Florida, *Transactions of the Gulf Coast Association of Geological Societies*, 18: 439–450.

Hunter, M.E., 1978, What is the Caloosahatchee Marl?, in *Hydrogeology of south-central Florida*, Southeastern Geological Society Publication 20, pp. 61–88.

Johnson, L.C., 1888, The structure of Florida, *American Journal of Science*, (Series 3), 36: 230–236.

Jones, D.S., MacFadden, B.J., Webb, S.D., Mueller, P.A., Hodell, D.A., and Cronin, T.M., 1991, Integrated geochronology of a classic pliocene fossil site in Florida: linking marine and terrestrial biochronologies, *The Journal of Geology*, 99(5): 637–648.

Ketcher, K.M., 1992, Stratigraphy and Environment of Bed 11 of the "Pinecrest" Beds at Sarasota, Florida, in Scott, T.M. and Allmon, W.D., Eds., *Plio-Pleistocene Stratigraphy and Paleontology of Southern Florida*, Special Publication 36, Florida Geological Survey, Florida Department of Natural Resources, Tallahassee, FL, pp. 167–176.

King, K.C. and Wright, R., 1979, Revision of the Tampa Formation, west central Florida, *Transactions of the Gulf Coast Association of Geological Societies*, 29: 257–262.

Klitgord, K.D., Popenoe, P., and Schouten, H., 1984, Florida: a Jurassic transform plate boundary, *Journal of Geophysical Research*, 89: 7753–7772.

Krantz, D.E., 1991, A chronology of Pliocene sea level fluctuations, U.S. Atlantic Coastal Plain, *Quaternary Science Reviews*, 10: 163–174.

Lyons, W.G., 1991, Post-Miocene species of *Latirus* Montfort, 1810 (Mollusca: Fasciolariidae) of southern Florida, with a review of regional marine biostratigraphy, *Bulletin of the Florida Museum of Natural History*, 35(3): 131–208.

Lyons, W.G., 1992, Caloosahatchee-age and younger molluscan assemblages at APAC mine, Sarasota County, Florida, in Scott, T.M. and Allmon, W.D., Eds., *The Plio-Pleistocene Stratigraphy and Paleontology of Southern Florida*, Special Publication 36, Florida Geological Survey, Florida Department of Natural Resources, Tallahassee, FL, pp. 133–159.

MacFadden, B.J., 1995, Magnetic polarity stratigraphy and correlation of the Leisey Shell Pits, Hillsborough County, Florida, in Hulbert, R.C., Morgan, G.S., and Webb, S.D., Eds., Paleontology and Geology of the Leisey Shell Pits, Early Pleistocene of Florida, Part 1, *Bulletin of the Florida Museum of Natural History*, 37, Part 1(1–10): 107–116.

Mansfield, W.C., 1930, Miocene Gastropods and Scaphopods of the Choctawhatchee Formation of Florida, Geological Bulletin 3, Florida Geological Survey, Florida Department of Conservation, Tallahassee, FL, 142 pp.

Mansfield, W.C., 1931, Some Tertiary mollusks from southern Florida, *Proceedings of the U.S. National Museum*, 79(21): 1–12.

Mansfield, W.C., 1932, Pliocene fossils from limestone in southern Florida, *U.S. Geological Survey Professional Paper*, 170: 43–56.

Mansfield, W.C., 1937, Mollusks of the Tampa and Suwannee Limestones of Florida, Bulletin 15, Florida Geological Survey, Florida Department of Conservation, Tallahassee, FL, 334 pp.

Mansfield, W.C., 1939, Notes on the Upper Tertiary and Pleistocene Mollusks of Peninsular Florida, Bulletin 18, Florida Geological Survey, Department of Conservation, Tallahassee, FL, 75 pp.

Matson, G.C. and Clapp, F.G., 1909, A Preliminary Report on the Geology of Florida, with Special Reference to the Stratigraphy, Florida Geological Survey Annual Report 2, Florida Geological Survey, Tallahassee, FL, pp. 25–173.

Maurasse, F. and Glass, B., 1976, Radiolarian stratigraphy and North American microtektites in Caribbean RC8–58: implications concerning Late Eocene radiolarian chronology and the age of the Eocene-Oligocene boundary, in Causse, R., Ed., *Transactions of the 7th Caribbean Geological Conference, Guadeloupe*, pp. 205–212.

McCaslin, J.C., April 28, 1986, NW Florida produces most of the state's oil, *Oil and Gas Journal*, 103–104.

McGinty, T.L., 1940, New land and marine Tertiary shells from southern Florida, *The Nautilus*, 53(3): 81–84.

McGinty, T.L., 1970, Mollusca of the "Glades" Unit of southern Florida: Part 1. Introduction and Observations, *Tulane Studies in Geology and Paleontology*, 8(2): 53–56.

Meeder, J.F., 1979, *A Field Guide with Road Log to the Pliocene Fossil Reef of Southwest Florida*, Field Trip Guide (1979), Miami Geological Society, Coral Gables, FL, 19 pp.

Meeder, J.F., 1980, New information on Pliocene reef limestones and associated facies in Collier and Lee counties, Florida, in Gleason, P., Ed., *Water, Oil, and the Geology of Collier and Lee Counties*, Miami Geological Society, Coral Gables, FL, pp. 27–30.

Merritt, M.I., Myer, F.W., Sonntag, W.H., and Fitzpatrick, D.J., 1983, Subsurface Storage of Freshwater in South Florida: A Prospectus, U.S. Geological Survey Water Resources Investigations Report 83–4214, pp. 29–40.

Miller, W.L., 1987, Lithology and the Base of the Surficial Aquifer System, Palm Beach County, Florida, U.S. Geological Survey Water Resources Investigations Report 86–4067, 1 map.

Missimer, T.M., 1984, The geology of South Florida: a summary, in Gleason, P.J., Ed., *Environments of South Florida, Present and Past 2*, Miami Geological Society Memoir 2, Miami Geological Society, Coral Gables, FL, pp. 385–404.

Missimer, T.M., 1992, Stratigraphic relationships of sediment facies in the Tamiami Formation of southwest Florida: proposed intraformational correlations, in Scott, T.M. and Allmon, W.D., Eds., *The Plio-Pleistocene Stratigraphy and Paleontology of Southern Florida*, Special Publication 36, Florida Geological Survey, Florida Department of Natural Resources, Tallahassee, FL, pp. 63–92.

Morgan, G.S. and Hulbert, R.C., 1995, Overview of the geology and vertebrate biochronology of the Leisey Shell Pit Local Fauna, Hillsborough County, Florida, in Hulbert, R.C., Morgan, G.S., and Webb, S.D., Eds., Paleontology and Geology of the Leisey Shell Pits, Early Pleistocene of Florida, Part 1, *Bulletin of the Florida Museum of Natural History*, 37, Part 1(1–10): 1–92.

Mossom, D.S., 1925, A Preliminary Report on the Limestones and Marls of Florida, Florida Geological Survey Annual Report 16, Florida Geological Survey, Tallahassee, FL, pp. 27–203.

Olsson, A.A., 1967, *Some Tertiary Mollusks from South Florida and the Caribbean*, Paleontological Research Institute, Ithaca, NY, 61 pp.

Olsson, A.A. and Harbison, A., 1953, *Pliocene Mollusca of Southern Florida, with special reference to those from North Saint Petersburg*, Monographs of the Academy of Natural Sciences of Philadelphia, 8: 1–361.

Olsson, A.A. and Petit, R.E., 1964, Some Neogene Mollusca from Florida and the Carolinas, *Bulletins of American Paleontology*, 47(217): 509–575.

Olsson, A.A. and Petit, R.E., 1968, Notes on *Siphocypraea*, *Bulletins of American Paleontology*, 54(242): 279–289.

Osmond, J.K., Carpenter, J.R., and Windom, H.L., 1965, Th 230/U 234 age of the Pleistocene corals and oölites of Florida, *Journal of Geophysical Research*, 70: 1843–1847.

Parker, G.G. and Cooke, C.W., 1944, Late Cenozoic Geology of Southern Florida, with a Discussion of the Ground Water, Florida Geological Survey Bulletin 27, Florida Geologic Survey, Florida Department of Conservation, Tallahassee, FL, 119 pp.

Parker, G.G., Ferguson, G.E., and Love, S.K., 1955, Water Resources of Southeastern Florida, U.S. Geological Survey Water-Supply Paper 1255, 965 pp.

Parodiz, J.J., 1988, A new species of *Siphocypraea* (Gastropoda:Cypraeidae) from the Neogene of southwest Florida, *Annals of the Carnegie Museum*, 57(3): 91–97.

Parodiz, J.J. and J.J. Tripp, 1992, The Neogene Cassidae of southern Florida, with description of a new species of *Cassis* (Gastropoda:Prosobranchia), *Annals of the Carnegie Museum*, 61(4): 317–325.

Petuch, E.J., 1976, An unusual molluscan assemblage from Venezuela, *The Veliger*, 18: 320–325.

Petuch, E.J., 1979, A new species of *Siphocypraea* from northern South America, with notes on the genus in the Caribbean, *Bulletin of Marine Science*, 29(2): 216–225.

Petuch, E.J., 1982, Notes on the Molluscan Paleontology of the Pinecrest Beds at Sarasota, Florida, with the description of *Pyruella*, a stratigraphically important new genus, *Proceedings of the Academy of Natural Sciences of Philadelphia,* 134: 12–30.

Petuch, E.J., 1986, The Pliocene reefs of Miami: their geomorphological significance in the evolution of the Atlantic coastal ridge, southeastern Florida, U.S., *Journal of Coastal Research,* 2(4): 391–408.

Petuch, E.J., 1987a, The Florida Everglades: a buried pseudoatoll? *Journal of Coastal Research,* 3(2): 189–200.

Petuch, E.J., 1987b, A new *Ecphora* fauna from southern Florida, *The Nautilus,* 10(4): 200–206.

Petuch, E.J., 1988a, *Neogene History of Tropical American Mollusks: Biogeography and Evolutionary Patterns of Tropical Western Atlantic Mollusca,* The Coastal Education and Research Foundation, Charlottesville, VA, 217 pp.

Petuch, E.J., 1988b, A new giant *Muricanthus* from the Pliocene of southern Florida, *Bulletin of Paleomalacology,* 1(1): 17–21.

Petuch, E.J., 1989a, *Field Guide to the Ecphoras,* The Coastal Education and Research Foundation, Charlottesville, VA, 140 pp.

Petuch, E.J., 1989b, New species of *Malea* from the Pleistocene of southern Florida, *The Nautilus,* 103(3): 92–95.

Petuch, E.J., 1990, New gastropods from the Bermont formation (middle Pleistocene) of the Everglades Basin, *The Nautilus,* 104(3): 96–104.

Petuch, E.J., 1991, *New Gastropods from the Plio-Pleistocene of Southwestern Florida and the Everglades Basin,* Special Publication 1, W.H. Dall Paleontological Research Center, Florida Atlantic University, 85 pp. (copies available at the Smithsonian Institution and the Carnegie Museum).

Petuch, E.J., 1993, Patterns of Gastropod Extinction in the Plio-Pleistocene Okeechobean Sea of Southern Florida, in Zullo, V. et al., Eds., The Neogene of Florida and Adjacent Regions, *Proceedings of the Third Bald Head Island Conference on Coastal Plains Geology,* Special Publication 37, Florida Geological Survey, Tallahassee, FL, pp. 73–85.

Petuch, E.J., 1994, *Atlas of Florida Fossil Shells (Pliocene and Pleistocene Marine Gastropods),* The Graves Museum of Archaeology and Natural History, Dania, FL, and Spectrum Press, Chicago, IL, 394 pp.

Petuch, E.J., October 13, 1995, Molluscan diversity in the Late Neogene of Florida: evidence for a two-staged mass extinction, *Science,* 270, pp. 275–277.

Petuch, E.J., 1996, *Calusacypraea,* a new, possibly Neotenic, genus of cowries from the Pliocene of southern Florida, *The Nautilus,* 110(1): 17–21.

Petuch, E.J., 1997, A new gastropod fauna from an Oligocene back-reef lagoonal environment in west central Florida, *The Nautilus,* 110(4): 122–138.

Petuch, E.J., 1998, New cowries from the Myakka Lagoon System and the Everglades Pseudoatoll (Pliocene Okeechobean Sea). *La Conchiglia,* 30(288): 27–37.

Petuch, E.J., 2004, *Cenozoic Seas: The View from Eastern North America,* CRC Press, Boca Raton, FL, 308 pp.

Pilsbry, H.A., 1934, Review of the Planorbidae of Florida, with notes on other members of the family, *Proceedings of the Academy of Natural Sciences of Philadelphia,* 86: 29–66.

Poag, C.W., 1999, *Chesapeake Invader: Discovering America's Giant Meteor Crater,* Princeton University Press, NJ, 183 pp.

Poag, C.W., Koeberl, C., and Reimold, W.U., 2004, *The Chesapeake Bay Crater: Geology and Geophysics of a Late Eocene Submarine Impact Structure,* Springer-Verlag, New York, 522 pp.

Portell, R.W., Schindler, K.S., and Nicol, D., 1995, Biostratigraphy and Paleoecology of the Pleistocene Invertebrates from the Leisey Shell Pits, Hillsborough County, Florida, in Hulbert, R.C., Morgan, G.S., and Webb, S.D., Eds., Paleontology and Geology of the Leisey Shell Pits, Early Pleistocene of Florida, Part 1, *Bulletin of the Florida Museum of Natural History,* 37, Part 1(1–10): 127–149.

Powars, D.S., 2000, The effects of the Chesapeake Bay impact crater on the geologic framework and the correlation of hydrogeologic units of Southeastern Virginia, south of the James River, U.S. Geological Survey Professional Paper 1622, U.S. Geological Survey, 53 pp.

Powars, D.S. and Bruce, T.S., 1999, The effects of the Chesapeake Bay impact crater on the geological framework and correlation of hydrogeologic units of the Lower York-James Peninsula, VA, U.S. Geological Survey Professional Paper 1612, U.S. Geological Survey, 82 pp.

Puri, H.S. and Vernon, R.O., 1964, *Summary of the Geology of Florida,* Special Publication 5, Florida Geological Survey, Florida Department of Natural Resources, Tallahassee, FL, 312 pp.

Puri, H.S. and Winston, G.O., 1974, *Geologic Framework of the High Transmissivity Zones in South Florida,* Special Publication 20, Bureau of Geology, Florida Department of Natural Resources, Tallahassee, FL, pp. 41–58, cross section C-D.

Randazzo, A.F. and Jones, D.S., Eds., 1997, *The Geology of Florida,* University Press of Florida, Gainesville, FL, 327 pp.

Sanfilipo, A.L., Riedel, W.R., Glass, B.P., and Kyte, F.T., 1985, Late Eocene microtectites and radiolarian extinctions on Barbados, *Nature,* 314: 613–615.

Sanford, S., 1909, The Topography and Geology of Southern Florida, Florida Geological Survey Annual Report 2, Florida Geological Survey, Tallahassee, FL, pp. 175–231.

Sanford, S., 1913, Geology and Ground Waters of Florida: Southern Florida, U.S. Geological Survey Water-Supply Paper 319, pp. 42–64.

Scott, T.M., 1988, *The Lithostratigraphy of the Hawthorn Group (Miocene) of Florida,* Bulletin 59, Florida Geological Survey, Florida Department of Natural Resources, Tallahassee, FL, 148 pp.

Scott, T.M., 1992, Coastal Plains Stratigraphy: the dichotomy of biostratigraphy and lithostratigraphy — a philosophical approach to an old problem, in Scott, T.M. and Allmon, W.D., Eds., *The Plio-Pleistocene Stratigraphy and Paleontology of Southern Florida,* Special Publication 36, Florida Geological Survey, Florida Department of Natural Resources, Tallahassee, FL, pp. 21–25.

Scott, T.M., 1997, Miocene to holocene history of Florida, in Randazzo, A.F. and Jones, D.S., Eds., *The Geology of Florida,* University Press of Florida, Gainesville, FL, pp. 57–67.

Sellards, E.H., 1919, Geologic Sections across the Everglades of Florida, Florida Geological Survey Annual Report 12, Florida Geological Survey, Tallahassee, FL, pp. 105–141.

Smith, M., 1936, New tertiary shells from Florida, *The Nautilus,* 49(4): 135–139.

Stubbs, S.A., 1940, Pliocene mollusks from a well at Sanford, Florida, *Journal of Paleontology,* 14(5): 510–514.

Swayze, L.J. and Miller, W.L., 1984, Hydrogeology of a Zone of Secondary Permeability in the Surficial Aquifer of Eastern Palm Beach County, Florida, U.S. Geological Survey Water-Resources Investigations Report 83-4249 (in cooperation with Palm Beach County), U.S. Geological Survey, 38 pp.

Tucker, H.I. and Wilson, D., 1932, Some new or otherwise interesting fossils from the Florida Tertiary, *Bulletin of American Paleontology,* 18(65): 1–24.

Tucker, H.I. and Wilson, D., 1933, A second contribution to the Neogene paleontology of South Florida, *Bulletin of American Paleontology,* 18(66): 1–21.

Vail, P.R. and Hardenbol, J., 1979, Sea-level changes during the Tertiary, *Oceanus,* 23(3): 71–79.

Vail, P.R., Mitchum, R.M., and Thompson, S., 1977, Global cycles of sea level changes, in Payton, C.E., Ed., *Seismic Stratigraphy: Applications to Hydrocarbon Exploration,* American Association of Petroleum Geologists Memoir 26, pp. 83–97.

Vaughan, T.W., 1919, Fossil corals from Central America, Cuba, and Porto Rico, with an account of the American Tertiary, Pleistocene, and Recent coral reefs, *U.S. National Museum Bulletin,* 103: 189–524.

Vermeij, G.J. and Vokes, E.H., 1997, Cenozoic Muricidae of the western Atlantic Region. Part 12: The Subfamily Ocenebrinae (in part), *Tulane Studies in Geology and Paleontology,* 29(3): 69–18.

Vokes, E.H., 1963, Cenozoic Muricidae of the western Atlantic Region. Part 1: *Murex* sensu stricto, *Tulane Studies in Geology,* 1(3): 93–123.

Vokes, E.H., 1965, Cenozoic Muricidae of the western Atlantic Region. Part 2: *Chicoreus* sensu stricto and *Chicoreus (Siratus), Tulane Studies in Geology,* 3(4): 181–204.

Vokes, E.H., 1966, The genus *Vasum* (Mollusca: Gastropoda) in the New World, *Tulane Studies in Geology,* 5(1): 1–36.

Vokes, E.H., 1975, Cenozoic Muricidae of the western Atlantic Region. Part 6: *Aspella* and *Dermomurex, Tulane Studies in Geology and Paleontology,* 11(3): 121–162.

Vokes, E.H., 1976, Cenozoic Muricidae of the western Atlantic Region. Part 7: *Calotrophon* and *Attiliosa, Tulane Studies in Geology and Paleontology,* 12(3): 101–136.

Waldrop, J.S. and Wilson, D., 1990, Late Cenozoic stratigraphy of the Sarasota area, in Allmon, W.D. and Scott, T.M., Eds., *Plio-Peistocene Stratigraphy of South Florida,* Southeastern Geological Society Annual Fieldtrip Guidebook 31, 221 pp.

Ward, L.W., 1992, Diagnostic Mollusks from the APAC Pit, Sarasota, Florida, in Scott, T.M. and Allmon, W.D., Eds., *Plio-Pleistocene Stratigraphy and Paleontology of Southern Florida,* Special Publication 36, Florida Geological Survey, Florida Department of Natural Resources, Tallahassee, FL, pp. 161–165.

Weisbord, N.E., 1974, Late Cenozoic Corals of South Florida, *Bulletins of American Paleontology,* 66(285): 259–544.

White, W.A., 1970, The Geomorphology of the Florida Peninsula, Florida Geological Survey Bulletin 51, Tallahassee, FL, pp. 1–185.

Index

A

Acline Fauna, 72
Aftonian Pleistocene era, 19, 147
Akleistostoma genus, 14
Alligator Alley, 2–3, 72
Alligator teeth, in Okeelanta Member, 166
Anastasia Formation, 11, 15
    age and correlative units, 192
    history of discovery, 189–190
    index fossils, 192
    lithologic description and areal extent, 190–191
    outcrops at Jupiter Blowing Rocks, 190
    stratotype, 192
    view of blow-hole, 191
APAC quarry, 2
Apalachee Gyre, 42
Aquitanian Miocene era, 17
    simulated space shuttle image, 33
Arcadia Formation, 8, 9
    age and correlative units, 34
    areal distribution, 43
    Hawthorne Group, 32–33
    lithologic description and areal extent, 32–34
    Nocatee Member, 34
    in Paleogene era, 32–33
    Tampa Member, 36–41
Arcadia Subsea, 17, 18
    Paleogene era paleogeography, 41–43
Areal extent
    Anastasia Formation, 190–191
    Arcadia Formation, Hawthorn Group, 32–34, 43
    Ayers Landing Member, 134
    Bee Branch Member, 131–132
    Belle Glade Member, 168–169
    Bermont Formation, 150–152
    Buckingham Member, 76, 77
    Caloosahatchee Formation, 119–120
    Coffee Mill Hammock Member, 180–182
    Fordville Member, 121–122
    Fort Denaud Member, 124, 125
    Fort Drum Member, 140–141
    Fort Thompson Formation, 176–177
    Fruitville Member, 104–106
    Golden Gate Member, 96
    Holey Land Member, 154–156
    Key Largo Formation, 192–193
    Lake Flint Formation, 194–195
    Miami Formation, 187–188
    Nashua Formation, 137–138
    Nocatee and Tampa Members, 35
    Ochopee Member, 92–93
    Okaloacoochee Member, 177, 178
    Okeelanta Member, 162–164
    Pamlico Formation, 195–197
    Pinecrest Member, 83–84
    Rucks Pit Member, 142
    Suwannee Formation, 28–29
    Tamiami Formation, 74
    Tampa Member, 36–37
Arikareean Land Mammal Age, 12
Ashland Petroleum and Asphalt Corp., 3
Asteroid impact, and Everglades unconformity, 22–25
Astroblemes, 22, 23
Atlantic Coastal Ridge, 4
Avon Park Formation, 9, 21, 25
Ayers Landing Member, 10
    age and correlative units, 135
    history of discovery, 134
    index fossils, 135–137
    lithologic description and areal extent, 134
    stratotype, 134–135

B

Back Reef, 97
    Cnidaria-Scleractinia of, 98
Bayshore Formation, 8, 9, 198
Bee Branch Member, 10
    age and correlative units, 132
    history of discovery, 131
    index fossils, 132–134
    lithologic description and areal extent, 131–132
    stratotype, 132
Belgrade Formation of North Carolina, 37
Belle Glade Member, 11
    age and correlative units, 169
    history of discovery, 168
    index fossils, 169–172
    lithologic description and areal extent, 168–169
    stratotype, 169
    view at Palm Beach Aggregates, Inc. quarries, 155
Belle Glade Subsea, 17, 19, 32, 147
    Belle Glade Member, 168–172
    paleogeography, 166–167
    simulated space shuttle image, 167
Bellegladean Mollusk Age, 12, 15
Bermont Formation, 8, 11, 32, 147, 198
    age and correlative units, 152
    history of discovery, 150
    Holey Land Member, 153–162
    index fossils, 152–153
    lithologic description and areal extent, 150–152
    in Middle Pleistocene era, 150–166
    Okeelanta Member, 162–166
    stratotype, 152
Bermont Lagoon System, 150, 167